THE OLD GUARD

&

The Old Guard Association

THE OLD GUARD

Turner®
Publishing Company

Mission Statement

To establish a permanent organization that will further develop the common bond existing between all who served with "The Old Guard."

To maintain records and publish the history and achievements of the 3rd Infantry Regiment and The Old Guard Association.

To be of assistance and service in matters pertaining to Veterans of "The Old Guard."

To perpetuate the memory of deceased Veterans of the 3rd Infantry Regiment "The Old Guard."

To maintain close liaison with the active duty members of the 3rd Infantry Regiment "The Old Guard" currently serving in the National Capital Region.

To recognize, by appropriate presentations, the Noncommissioned Officer and the Soldier-of-the-Year of the Regiment.

To carry out aims and purposes, The Association shall publish the Association's news letter, "PRIDE."

Turner Publishing Company
The Front Line of Military History
412 Broadway • P.O. Box 3101
Paducah, KY 42002-3101
(270) 443-0121
www.turnerpublishing.com
Turner Publishing Company Staff
Greg Wurth, Editor
Frene Melton, Designer

Library of Congress Cataloging-in-Publication Data

ISBN: 1-59652-008-6
Library of Congress Control No.: 2005926914

Printed in the United States of America.

0 9 8 7 6 5 4 3 2 1

Right: Ted Manolukas walking post, 1951.

Contents

Executives' Messages

Third Infantry Regiment "The Old Guard" passes in review, Parade Grounds, Fort Myer, VA, 1955. Color Bearers (left to right): Pvt. W.P. Madden, unidentified, Pvt. H.J. Andrews, unidentified.

The Old Guard Association
P.O. Box 1785 Fort Myer, VA 22211

December 2004

Fellow Old Guard Soldiers:

Welcome to the premier edition of the Old Guard and Old Guard Association History and Yearbook. On behalf of the Officers and Staff of both organizations, it is our pleasure to publish this book for the enjoyment of you, your family and friends.

Over the Regiment's long and storied past, many histories have been written and presented to the public. Some were official and authorized by the Department of the Army - others were collections of writings from various eras and sources. Similarly, many yearbooks have been compiled, usually containing only small photographs of Soldiers of a particular year, together with their names.

Now, for the first time in the Regiment's annals, we are pleased to present a comprehensive history of the Army's oldest Infantry Regiment, together with a multi-era yearbook in which you will find then-and-now photographs and biographies of many of the men and women who have proudly worn the buff strap and cockades of The Old Guard.

The Old Guard Association and Turner Publishing Company wish to thank the U.S. Army Center for Military History, The Old Guard Museum, The Old Guard Public Affairs Office and the many present and former Soldiers of the Regiment who dug down in their footlockers, duffel bags and memories, in search of much of the historical information, stories, anecdotes and photographs in this volume.

Many years from now, perhaps your children or their children, will come across this book long after you have stood Final Muster, and they will learn a bit more about your honored service to this Nation. They will smile as they recall the small stories you told about your pulling a 'fast one' on your First Sergeant, or a practical joke on a Buddy. They will gaze upon the photo of a fresh-faced young man or woman being "all they can be".

Whatever the occasion someone has to open this book in the future, you can be sure that it will speak highly of you, your Regiment, your Army and your Association.

Now, join me as we return with our forebears to Lundy's Lane and Chapultepec, Gettysburg, Luzon, Newfoundland and Berlin... to the frozen mountains of Korea and the steaming jungles of Vietnam... to Djibouti, Ethiopia and Iraq.

Sit back, relax, and recall with PRIDE your storied and historic past, for you are an Old Guardsman!

Fraternally,

Donald G. Cofsky
Executive Editor

The Old Guard Association

P.O. Box 1785 Fort Myer, VA 22211

December 2004

Fellow Association Members:

As President of The Old Guard Association, I am very pleased to see this history of our Regiment and our Association published to commemorate our two great organizations and those who serve them.

In this 220th anniversary year of our Regiment, and the seventh anniversary of our Association, we continue to work together in support of our mutual goals and objectives, to provide a link between past and present members of The Old Guard.

Our membership spans service to our country in World War II, Korea, Vietnam, Cold War "hot spots', Grenada, Panama, Desert Shield/Storm, Somalia, Haiti, the former Yugoslavia, Bosnia and Kosovo, and currently in support of Global War on Terrorism operations in Iraq, Afghanistan, and the Horn of Africa. As you read the Regimental history and the biographies of our members, you see what service to our great country is all about.

A quote from one of John Ford's classic films, *FORT APACHE*, stands out in my mind as a fitting description of what TOGA and the 3d U.S. Infantry are all about. The scene shows John Wayne, as the Regimental Commander, discussing the upcoming campaign and commenting on a reporter's question about the forgotten members of the Regiment. He replies, "They are still alive. The faces and names may change, but they remain - the Regiment, the Regular Army". Our faces and names will change in the future, but the Regiment and TOGA will remain.

Enjoy this book now and in the future, and continue to enjoy your affiliation with our Regiment and our Association.

Noli Me Tangere,

Pete McDermott

Pete McDermott, President
Old Guardsman, 1960-1962

Headquarters
United States Army Military
District of Washington
103 Third Avenue
Fort Lesley J. McNair, DC 20319-5058
04 August 2004

To the Officers and Soldiers of The Old Guard:

It is my distinct pleasure to take this opportunity to directly address the officers and soldiers of this remarkable Regiment, The Old Guard of the Army.

Since my arrival last year, I have observed you closely during the countless ceremonies, memorial affairs, and executing contingency plans and operations, such as Joint Task Force-Horn of Africa and at the Reagan Statue Funeral. The performance throughout has been nothing but spectacular. Your performance reflects the commitment to outstanding professional service to the United States Army and the Military District of Washington.

Thank you for all you do on a daily basis, reinforcing that Soldiers are the centerpiece of the Army and our Nation.

Sincerely,

Galen B. Jackman
Major General, U.S. Army
Commanding

DEPARTMENT OF THE ARMY
3d U.S. INFANTRY REGIMENT (THE OLD GUARD)
FORT MYER, VA 22211

REPLY TO
ATTENTION OF

ANOG-CD

June 30, 2004

To the Officers, Non-Commissioned Officers and Soldiers of The Old Guard:

During this historic year, the soldiers of The Old Guard have added much to our proud lineage with accomplishments ranging from a successful combat deployment to the President Reagan State Funeral. Together we have sustained a great team, united in our effort to meet the many ceremonial, memorial and special security missions required of this Regiment to support the Army and the Nation. I am proud to have stood with you on the fields of honor and on tactical missions. You can be justifiably proud of all you have accomplished.

Our magnificent soldiers of this Regiment have built upon the layered standards of excellence from previous generations of Old Guard Soldiers. No unit in the Army is expected to meet the stringent standards of ceremonial excellence, while maintaining tactical skills to defend our Nation. This past year solidified the level of training necessary to achieve and sustain our operationally relevant and ready force. Through numerous ceremonies and tactical external evaluations and mission execution, the Regiment proved that whether in the Horn of Africa or Summerall Field, discipline and precision is The Old Guard standard.

I also wish to extend my personal thanks to the members of The Old Guard Association (TOGA). Their support and dedication to the Soldiers of the Regiment remains unwavering. They form the unbroken chain of excellence which connects past, present, and extends to the future. The efforts of TOGA to record and publish the history of this great Regiment deserve the respect and admiration of every soldier.

For the superb fashion you have accomplished every task, for the hours of arduous training, for the unwavering, loyal support you have given – you shall always command my personal devotion and admiration. To each of you I say, "continue the mission with excellence." The Army and the Nation will never forget all that you have done.

Charles L. Taylor
Colonel, Infantry
Commanding

THE OLD GUARD

The History of The Old Guard and Old Guard Association

History of the Old Guard

Caissons with 75mm gun.

The Early Years

The U.S. Army's oldest regiment, The Old Guard, was created as a result of the 1783 Peace Treaty of Paris. Among the provisions of this treaty ending the war between Britain, France, and the colonies of British America (Americans know the war as the American Revolution) was the requirement that the newly independent colonies take military control and civil responsibility for the area of land west to the Mississippi River. Native Americans and their British allies inhabited this land.

The American army that had won the Revolution (with the help of a French army and French fleet) had been disbanded and the troops returned to their respective states in the spring of 1784. The Commander-in-Chief had bid good-bye to his officers, and returned to his farm on the Potomac in Virginia. A single artillery unit, posted to West Point, was retained from the Continental Army. For practical purposes, there was no force left to defend the United States. Congress was forced, because of the provisions of the treaty, to create an army.

The single unit created became the Old Guard. The troops of the new unit, probably not more than 450 men, and officers, never more than a couple dozen in the first years, were the last of the troops from Connecticut, New Jersey and Pennsylvania who were induced to enlist (having first been mustered out of the old army.) The officers held their commissions from the states from which they came, since there was no legislation authorizing the Confederation of States to issue commissions. These were the beginnings of the national defense and the United States Army.

The Treaty of Paris required that the United States accomplish three missions. First, to receive the garrisons west of the Ohio River that were held by the British Army, and to garrison some of those forts. Second, to control the flow of settlers from the east, a large number of who were claiming lands offered to them in lieu of Continental Army pay long in arrears. And lastly, to attempt to control the inevitable clashes between the Indians and the settlers and land speculators. The force that was created to accomplish these missions was called the 1st American Regiment. The 1st American Regiment began to train at a series of forts along the Pennsylvania frontier in the fall of 1784. The first commanding officer, Lieutenant Colonel Josiah Harmar, began to plan a campaign to

General Anthony Wayne. Courtesy of the Library of Congress.

subdue the Indians and then force the surrender of the British garrisons. His efforts in the next years—and those of other commanders in later years—met with only limited success (and occasionally with catastrophic failure), largely because the weak national government could not adequately support sustained military operations on the frontiers.

For the first 10 years of the existence of the Regiment, it fought under several names. Created as the 1st American Regiment, it was variously known as the American Regiment, the 1st Infantry, and the 1st Sub-Legion of the Legion of the United States. Under this last name, as a combined arms organization consisting of highly trained and specialized Infantry, Artillery, and Cavalry, the Old Guard became an elite regular infantry unit. The Legion, under the command of Pennsylvanian (and stern disciplinarian) Major General Anthony Wayne, finally broke the back of Indian resistance to settlement in the Old Northwest. One eventual outcome of The Battle of Fallen Timbers was that the British were finally forced to surrender Detroit, their last garrison in the United States, in 1796. When the mission of the Legion was accomplished, its parts were separated, and the Regiment again became the 1st Infantry.

A New Name

At the outset of War in America in 1812, the Regiment was sent to the Niagara frontier in New York and participated in the more or less continuous combat in that theater of war, including several attempts to invade British Canada. When peace was declared in 1815, it was largely because the powers of Europe were exhausted by 25 years of war with France and Napoleon. The United States Army returned to a peacetime establishment at the end of the conflict. That peacetime establishment required a force reduction (a "downsizing" in modern terms) of more than 30 regiments in the infantry alone. The forty-six regiments of infantry were, in 1815, consolidated into eight, mostly by combining the troops from 5 or more regiments into one unit and then renumbering the surviving organizations. The old (pre-1815) 1st Infantry was the oldest unit used to make up the new (post-1815) 3d Infantry, which created direct lineage as the army's oldest infantry unit. The commanders of the new units and their numbers were chosen on the basis of seniority. The first commander of the new 3d Infantry was Colonel John Miller, previously of the 28th Infantry and third in seniority in the Army. The number of the unit became the 3d as a result of Miller's seniority.

The westward expansion of the United States during the hundred years after 1784 was not peaceful. Old Guardsmen were engaged in the same exploration, peacemaking, peacekeeping, building, and combat missions as the rest of the army. The Old Guard was responsible for beginning to build many of the forts of the west, some of which became cities as settlement continued. Fort Dearborn eventually became a city of that name. Fort Howard contributed to the development of Green Bay, Wisconsin. Jefferson Barracks, Missouri was built in what became the city of St. Louis. In Kansas, the regimental commander, Colonel Henry Leavenworth, chose the site of the fort which bears his name. Camp Supply, Oklahoma, is now a state historic site. Many other posts, forts, or fortifications were built or manned by the Old Guard.

After the war of 1812, many of the traditional patterns of Army life in the west, and many of the traditional missions of the Regular Army in American life begin to be evident. Serving on the frontiers, the Army built most of its own posts, most sized for only a few companies operating independently of their regimental headquarters. As the only body of men under the control of the government with a clear mandate to use force to implement the national program of resettlement of the Indians, the Army was central to the policy of gaining military control while protecting the Indians from unscrupulous or terrified settlers. One of the most difficult of these situations concerned the Republic of Texas, its border with the United States, and the adjacent Indian Territory. Through the 1820's and 1830's, the Old Guard was stationed in Louisiana, Arkansas and Missouri, guarding in depth the border with the Texas Republic and the Indian Territory.

In April 1840 the entire 3d Infantry Regiment, numbering around 690 men, was sent to Florida to participate in the war against the Seminole Indians. The three year conflict ended by negotiation in time for the Regiment to be sent again to St. Louis as instructors and demonstration troops in the School for Brigade Drill at Jefferson Barracks, earning the first recorded nickname for the Regiment, the "buff sticks," after a flat stick of wood with a soft piece of leather attached that was commonly used to shine metal buttons and other uniform parts. After a year, the Regiment moved again to the Texas border as part of the "Army of Observation," protecting the U.S. border from the Texans and Mexicans. On the border, the regulars created "Camps of Instruction" to increase the level of combat readiness of the army as a whole. They were soon to need it.

War & Tradition

The war with Mexico, which began as a result of the admission of Texas to the union in 1846, is not well remembered today. It was, in its time, an unpopular war conducted by American armies invading Mexico from the north and east. Always outnumbered, operating without secure supply lines after amphibious landings in a foreign, armed, and hostile country, the American troops found their enemy entrenched and fortified in his own cities and towns, with weapons equal or superior in quality to their own. Most serious military observers expected the Americans to be beaten by the Mexican Army of Santa Anna. American supply lines were considered to be too long and difficult to maintain, and the overwhelming numbers of the Mexican Army appeared to be, before the war began, a telling precursor of disaster. The U.S. victory became an example used to teach and inspire in the American army for generations, and the war a training ground for the senior officers of our own Civil War. Success in the face of such overwhelming odds explains why the war has such a central part in the heraldry and traditions of the Old Guard and other regular regiments.

There were two major campaigns in the sixteen month war, and the 3d Infantry played a vital role in both of them. The northern invasion that began the war, after the Mexican attack on American forces at Palo Alto, Texas and Resaca de la Palma culminated in the attack and capture of the Mexican city of Monterey, 130 miles from the border. On the 21st of September 1846, the 3d Infantry, 262 men, was part of the first wave to attack the city. Enemy artillery was stationed at street intersections, and the narrow streets were bounded on each side with buildings filled with infantry. In penetrating the first line of defense of the walled city, the Regiment found itself drawn into a situation in which the limited space to maneuver made it difficult for American units to support each other. The losses that ensued reflect the nature of the combat. Over forty enlisted men were casualties; five officers, including the three senior officers, were killed and five wounded. The unit took its objective and held its ground until a lack of ammunition made withdrawal a necessity. Even then it remained under Mexican artillery fire during the night and into the morning, when the city fell to the American army. The bitter house to house defense of the city by the Mexican troops made it the most costly single day in the history of the 3d Infantry before Gettysburg. The anniversary of that day in battle was chosen in 1920 as the unit's Organization Day.

The southern campaign of the war began in December 1846, as in following months The Old Guard and other troops moved back to the United States and then embarked on transports. The 3d Infantry was transferred to a force under the command of Major

General Winfield Scott, whose campaign began formally on 9 March 1847 when General Scott's army participated in the first large scale amphibious landing ever undertaken by the U.S. Army, at Vera Cruz on the east coast of Mexico. The 3d Infantry, among the first ashore, helped to secure the city which was the anchor of the American supply line for the campaign on the Mexican capital, 200 miles inland. It is in this campaign that most of the traditions of the Regiment were born, enduring to be recognized officially almost a century later.

From Vera Cruz, the Army made straight for Mexico City. The Old Guard fought during the campaign as part of the brigade of regulars, the 1st Brigade of the 2d Division. The first resistance was at Cerro Gordo, a small, crossroads, overlooked by hills. The fight for the town would create another Regimental tradition.

The heights of Cerro Gordo were occupied by Mexican artillery and though the Americans had artillery positions that could bombard the hill their only means of driving the enemy off was a direct assault. Common to the tactics of the day was a bayonet charge, undertaken as the best and fastest way to cover the ground to an objective and grapple decisively with an enemy.

Gen. Winfield Scott wrote of the assault that "... the style of execution which I had pleasure to witness, was brilliant and decisive. The brigade ascended ...without shelter, and under the tremendous fire of artillery and muskets ... (they) drove the enemy from (the breastworks).. and after some minutes of sharp firing, finished the conquest with the bayonet ..." A long tradition in European and American armies is a show of subservience to the civil authorities, recognized by marching in peacetime under arms without bayonets fixed—thereby not ready to attack or threaten—and to keep colors covered except in sight of the enemy in time of war. The 3d Infantry is one of the few units officially authorized to pass in review with bayonets fixed. The practice commemorates the bayonet assault that the Regiment undertook with the 7th Infantry up Telegraph Hill at Cerro Gordo, and was authorized by the War Department in 1922.

The battles of Cerro Gordo (18 April), Contreras (19-20August), and Churubusco (20 August) were fought on the march to the Mexican capital. Bayonet charges at Cerro Gordo and Churubusco taught that the Mexican troops could not hold against determined men properly supported. Mexico City, the objective, was a walled city, defended in depth by means of outlying fortifications and fortified buildings, some of which could be reached only by elevated causeways cut at intervals by heavily defended gates and other strong points. The last of these fortified places to be taken was Chapultepec, the military academy of Mexico. A fortress sited atop a steep hill, it could be approached only by a causeway defended by artillery and infantry. Further fortifications surrounded the fortress on level ground. In the final assault of the Brigade of Regulars the 3d Infantry stormed the walls, raising the American flag on the parapet and establishing a foothold in the fortress which could be exploited. This was accomplished with yet another bayonet charge which so demoralized the defenders of Chapultepec that most of the Mexican troops soon surrendered. One unit which did not, the Cadets of the academy, were surrounded. They fought to their deaths or threw themselves off the parapets, creating an enduring Mexican military tradition to parallel that of the 3d Infantry.

The next day, the 14th of September, the 3d Infantry had the honor of marching at the head of the brigade as the lead element in the review of American troops as it entered the Mexican capital. It was at this time that the Regiment is said to have received its greatest legacy of the war. The Army commander, Major General Scott, in what is perhaps the greatest praise he could have given, is said to have turned to his staff as the 3d Infantry passed and said, "Gentlemen, take off your hats to the Old Guard of the Army." His purported remark, first officially used eighty years later, gave the Regiment its modern (and now official) name.

The regimental collection contains a single artifact of the event, a bandmaster's baton presented to the Regiment by Brigadier General Persifor Smith, the commander of the brigade in which they fought. Made of wood from the flagpole in the plaza in Mexico City, and mounted in Mexican silver, a story is told that it was a replacement for one broken in the assault. The Chapultepec baton is the Regiment's most important symbol, exhibited proudly in the Old Guard Museum.

The war in Texas and New Mexico against the Navajo was the next duty of the Regiment, and ten years was spent at it. Only a few anecdotes of the period survive, but they provide a diverse look at the life of the men in the Regiment. The first known concert by the Regimental Band, in El Paso, Texas, was in 1850. The story is also told of the march of the Regiment from one post to another in Texas that was interrupted by flash flooding which, because it occurred at night, carried away the camp and drowned a member of the senior NCO staff. The only baggage recovered was a chest of muskets found 20 feet up in a tree about 20 miles further downstream. The Regimental Headquarters and several companies were attacked at night in the summer of 1860 at Fort Defiance, New Mexico Territory. In one of the few recorded assaults by Indians against a fortified post, men of the 3d Infantry, in small, half dressed groups, fought all night against repeated assaults and defeated the Navajo by dawn.

To Save the Union

The Civil War began in the summer of 1861. In the spring President Lincoln had called for troops to defend the capital and the union, and the regular army was expected to set the example for citizen soldiers. The Regiment left Fort Clark, Texas, and marched southeast to meet transports which would carry five companies and the headquarters north, and two companies to Florida to garrison Fort Pickens. Three companies were not able to complete their

A scene from the Battle of Second Bull Run. Courtesy of the Library of Congress.

evacuation of Texas, and were forced to surrender. It was not until the spring of 1862 that the companies from Florida joined the Regiment, and shortly after, the attrition and casualties of the first year of the war forced four companies to be disbanded and the men to be distributed to the remaining six. The Old Guard spent the rest of the war as a battalion of six companies. Many of the officers were detached and performed staff duty at brigade and division levels, setting an example in Army administration as the nation struggled to form an army. The senior officer, for a time, was Captain George Sykes of Maryland, Company K.

In July of 1861, the Union advance into Virginia was stopped at Bull Run as the two armies groped and floundered into each other. When the federal line broke in the afternoon, and the routed army began to flee toward Washington, a movement by the southerners threatened to cut off the retreat. George Sykes, (now Major) with five companies of the 3d Infantry, two companies of the 2d, and one company of the 8th Infantry marched his ad hoc battalion to a critical ridge near the road to Stone Bridge. For the only time in the war, the battalion formed a square and successfully defended the ridge and road against infantry, artillery and cavalry until all units of the fleeing army had crossed the bridge. The Battalion then retired in good order to Washington. Fewer than five hundred men had saved the Army and perhaps the union. When the President came to review the troops at the end of the month, the army commander pointed out the little Battalion and said to Lincoln that "these are the men who saved your army." Lincoln replied "Yes, I have heard of them."

There was no further combat by Old Guardsmen in Virginia until June of 1862. The federal movement up the Virginia Peninsula toward Richmond was repulsed in the Seven Days Battles.

At Gaines Mill, while holding the Federal right, the 3d Infantry was attacked by overwhelming forces and stood its ground to save the line from collapsing. The line was held, but at great cost to the Regiment, including the death of the regimental commander, Major Nathan Rossell, who was the last field grade officer to serve with the unit until the end of the war. The rise in the ground where the combat took place is now called Regular's Hill.

The Old Guard was in heavy combat during the battle of Second Bull Run, with the Old Guard deployed as skirmishers in Groveton, and then fell back to hold Henry House Hill. The Old Guard had little part to play in Antietam but was deployed at a potentially critical point. At Fredericksburg, in December, the Brigade of Regulars was used to hold the town below the heights and the 3d Infantry was posted in a tannery. Under continuous shelling, the Regulars were finally deployed in support of the retreat and were the last troops to return to Falmouth Heights on the north side of the river.

After the defeat at Fredericksburg, the Army's new commander took action to boost morale and foster esprit de corps by reorganizing the Army and giving each unit down to division level its own insignia, based on the insignia of its parent Army Corps (soldiers already wore a company letter and

Left: General Scott leading the 3d Infantry in the Battle at Vera Cruz. Courtesy of the Library of Congress.

a regimental number). The system MG Joseph Hooker devised figures prominently in the heraldry of the Old Guard. The white maltese cross was worn by all men of the 1st Brigade, 2d Division, of the Fifth Corps, known as the Regular Division. It is the origin of the three crosses on the Coat of Arms of the Old Guard.

At Chancellorsville in the spring of 1863, the initial success of the deployment to the battlefield was followed by an attack east along the Orange Turnpike spearheaded by the Old Guard and the Regular Division. The successful attack was not adequately supported, and the Division withdrew to the Chancellor house. As the day wore on, the Division was sent to secure all of the roads north of Chancellorsville to deny them to the enemy and ensure that the line of communications was kept open. When, after Jackson's attack, the line of communications became a line of retreat, the Army passed through the Fifth Corps elements and re-crossed the river. The Regulars fought a rear-guard action, slowing down the pursuing enemy until the Army could successfully return to the north side of the Rappahannock River.

George Sykes, until Chancellorsville the Division commander—the most common term for the Regular Division was "Sykes's Regulars"—was promoted to Fifth Corps Commander. He, along with two others (William Penrose and William Hoffman) were 3d Infantry officers before the war who achieved the rank of Major General. Both Sykes and Penrose had been company commanders in 1860; Hoffman, much older, had been the regimental commander. Command of the Regular Division fell to Romeyn Ayres for the remainder of the war.

The command of the Regiment changed again in the days before the campaign for Gettysburg. John Wilkins, senior Captain and commander since the death of Major Nathan Rossell at Gaines Mill, was ranked by Captain Henry Freedly, whose parole from his capture in Texas had expired, allowing him to return to active duty. The Regiment entered the campaign as combat ready as any in the division. To a man, they were experienced veterans. Its single shortfall was manpower. More than fifty percent of its officer positions were vacant, and, its six companies--with a authorized strength of 576—had fewer than 300 men present for duty. The casualties of the Regiment were never replaced, and losses from desertion were also significant.

The final combat of the war for the 3d Infantry would be Gettysburg. The campaign leading to Gettysburg was essentially a race between armies. When the troops finally blundered into each other—some of the Southerners simply looking for shoes—there ensued a series of holding actions until the main bodies of the armies could be brought up to the town and deployed. The Federal line was along a ridge in the shape of a fish hook called Cemetery Ridge. At the south

The Battle of Gettysburg, July 3, 1863. Courtesy of the Library of Congress.

Officers of the 3d Infantry, 1865.

end of the line were two hills, Little Round Top and Round Top. That end of the line faced a small valley—called Plum Run Valley—across which was a rocky outcropping called Devil's Den. North of that, behind a fence, was the Rose Farm, a wheatfield, and a road perpendicular to the Round Tops. It was the single most critical piece of land on the south end of the battlefield, because control of the hills determined who could control the battlefield.

The Regular Division, 3d Infantry included, reached Gettysburg at 0030 on July 2, the second day of the battle, having marched from Falmouth Heights, facing Fredericksburg, in 30 days (including detours to attempt to capture Confederate Colonel John Mosby) by way of the old battlefield at Manassas and passes to the Shenandoah Valley of Virginia. In the previous week there had been but one cooked meal. Reveille was at 0300. There was no breakfast. They formed at the right end of the Federal line—near Wolf's Hill—to defend against an assault by Ewell's Corps which did not occur. Withdrawn to the rear, the Regulars were held in reserve at the center of the Federal line until early afternoon.

During the day, the Third Corps had moved forward of the Cemetery Ridge line to gain and hold a road. The movement had not been coordinated, and by mid-afternoon the Corps formed a salient in the line. The attack of Confederate General Longstreet's Corps, beginning at 1600, was intended to take control of the Round Tops, smash into the exposed Third Corps line on the Wheatfield Road, and begin to roll up the Federal line on itself along Cemetery Ridge. The attack was stopped by a handful of infantry and artillery on Little Round Top and the Regular Division in Plum Run Valley and the Rose Farm.

The Regular Division received orders to move at 1700, and was sent running to Little Round Top, deploying along the north slope. At about 1800, the Division was ordered forward through Plum Run Valley to support the Third Corps troops being pushed out of the Rose Farm and the Wheatfield, and to form a strong point to hold the road. Once they had gained the stone wall, driving off its defenders, the lead elements of the Division and all that followed could not move forward without becoming intermingled with other Federal troops being forced from the field across their front. The regular brigades were ordered to lie down in their ranks behind the wall—extending back across to an open ridge—which provided some concealment, a smaller target, and helped to avoid casualties from stray rounds. But they were forced to endure Confederate fire (Benning's Georgians) from Devil's Den on their left. At 1830, the Regulars of the 2d Brigade moved across the wall in formation, colors flying, wheeled to their left, and began to attack the Confederate forces streaming out of the woods. 1st Brigade, including the Old Guard, held the high ground some yards behind the wall. The 3d US anchored the line with the Regiment's right flank open on the road at the right of the Brigade line.

The Regular's attack caused the Confederates to send reinforcements. When they came streaming down the road and into the wheatfield, flanking and engulfing the 2d Regular Brigade, the pressure at the wall and on the undefended road was too great. The two Regular Brigades were ordered to retire to their old position at the base of Little Round Top.

The Regulars had to fight their way out. The remains of the 2d Brigade passed through the first, and both formed into line of battle—under fire—dressed their lines, and began to slowly and grudgingly give up the ground they had crossed less than an hour before. The muddy, rock strewn ground between the hill and the base of Little Round Top was given up only slowly, as the Regulars delivered perfectly timed crashing volleys into the Southerner's disordered lines. The Confederate assault ground to a halt.

The 3d Infantry, having the most forward position of the 1st Brigade in the advance, had the most exposed in the retreat, and was the last to leave the field. When artillery support was brought up to the base of Little Round Top to help slow the Confederate advance and try to save the two Regular Brigades, some elements of the Regiment were caught in the open. Lieutenant John H. Page, who survived to become the regimental commander before his retirement from the army in 1910, wrote later that:

"As we were falling back, we saw the battery officers at the base of [Little Round Top] waving their hats for us to hurry up. We realized that they wished to use canister, so took up the double quick. As I was crossing the swampy ground, Captain Freedly...was shot in the leg, fell against me, and knocked me down. When I got the mud out of my eyes, I saw the artillery men waving their hats to lie low. I got behind a boulder with a number of my men when the battery opened. The Rebels came from all directions for the guns.... They waved their battle flags, a dozen

being just in front of me [i.e., in his line of retreat, between Page and the base of Little Round Top]. A number [of the enemy] were shot down where we were; they then retreated through the wheatfield and the woods."

The end of the Regular Division was described by an eyewitness:

"The Regulars fought with determined skill and bravely for nearly an hour, then reluctantly fell back as if on drill, but sharply and bravely contesting every foot of ground. These things I saw, and I am glad, as a volunteer, to bear tribute to the United States Regulars."

By the time the 3d Infantry reformed and it and the 1st Brigade watched the darkness fall, the men were able to cook their first hot meal in a week. The losses in killed and wounded would never be made up. Of 300 men present for duty in the Old Guard, including only 12 officers, 74 were killed, wounded, or missing, almost 25 percent of the unit. The Old Guard had lost two commanding officers in as many hours. The price paid to stop the assault of Longstreet's Corps was even higher in other Regular regiments. The real importance of the sacrifice was that the failure of the Confederate attack, blunted on the Regulars, induced the Confederates to attack the center of the Federal line the next day. Losses in Pickett's Confederate Division would be even higher than those of the regulars and the failure of Picketts attack ended The ended The battle the confederate loss any real Southern hope of winning the war.

The Regular Division, now little more than just a brigade in strength, was to get no rest. By the middle of August, the Regulars, including the 3d Infantry, were sent north to New York City.

The conscription of men for service in the Army in 1863 was a new event in American history. Begun just after Gettysburg, it was soon followed by protests everywhere and riots in Ohio, New Hampshire, Vermont, and New York. The worst of these took place in New York City. Regulars, not subject to the whims of state politics, were sent to quell the riots and enforce the law.

The 3d, with the rest of the 1st Brigade, moved into the lawless areas of the city still wearing the uniforms that they had worn since camped on Falmouth Heights in May. Several days of fighting occurred at barricades in the streets and at buildings fortified by the rioters. In the end, they were no match for the Regulars, who were neither inclined nor induced to show much mercy to those who refused to serve in the Army. Finally camped at Washington Park and issued new uniforms—and with the East River handy for bathing—their ragged appearance was soon improved, and they began to enjoy a well deserved rest.

The combat effectiveness of the Regular Division was gone, and it was reduced in size to a brigade, leaving a small number of regiments to garrison New York harbor. The rest, including the 3d Infantry, returned to Virginia, to be put on guard duty on the Orange and Alexandria Railroad, one of the supply routes for federal troops operating in the Shenandoah Valley of Virginia. In September, at the beginning of the winter of 1864, Captain Andrew Sheridan, Commanding the 3d Infantry, sent the following letter to the Adjutant General:

"I have the honor to apply for the relief of the Third US Infantry...from duty in the field and an assignment to some post where they can recuperate and if practicable, recruit. The Third Infantry, after the trouble in Texas in which three companies and many officers were taken prisoners, have been constantly in the field. Ten companies have been consolidated into Six, and the effective command now consists of eight officers (three of whom are now Field & Staff and five upon line duty) and one hundred and fifty-eight rifles."

The 3d was relieved from line duty in February; yet, still, their war was not over. Ordered to join the 4th Infantry as headquarters guard for the Army of the

LEFT: *The 3d Infantry and others witnessing the surrender of Robert E. Lee at Appomattoxch, Virginia on April 9, 1865. Courtesy of the Library of Congress.*

The 3d Infantry Company's Indian company, 1891.

Potomac, the unit served in that capacity for the last 3 months of the war, and was present at the surrender of Lee's army at Appomattox Court House. The arrangement as General Meade's guard was made by John Wilkins, commander of the 3d after the death of Major Rossell in 1862. Wilkins's is the first known written reference to the 3d Infantry as an "Old Guard." In the period of martial law that followed Lincoln's assassination, the 3d was retained for a while in Washington as Provost Guard, and was the lead element in the Grand Review of the Army for President Andrew Johnson. The Regiment that had saved the Army at Bull Run participated in the last act of the war.

An Asian Jungle War

In 1899, the 3d Infantry embarked for the Philippines, taking a ship of the new US Army Transport Service through the Mediterranean and the new Suez Canal, arriving in Manila. Engaged in combat within three days, they were to stay in the Philippines for three years.

As a result of the war with Spain, the United States had inherited control of the former Spanish colonies of Cuba and Puerto Rico in the Caribbean and the Philippine Islands in the Western Pacific. In the latter, there had been combat against the Spanish, and the Philippine "Insurrectos" uniformed and officered in units modeled on European armies the "insurrectos," had been a significant contributor to the defeat of the Spanish in their country. They intended, as a result of their victory, to set up an independent, sovereign Philippine government. Aside from the strategic issues of supply depots in the western Pacific for an American Navy dependent on coal, most Americans believed that they should govern the islands so recently paid for in American blood, and the United States took action to set up a colonial administration. The result was war between the United States and the Philippine nationalists for control of the capital (Manila) and the surrounding area. The US Army, regulars and volunteers, became involved in its first Asian jungle war.

The nature of the war was different in context but not in practice from the duty the Regiment had seen in Kansas thirty years earlier. Patrolling out of garrisons in the larger towns, and manning outposts in smaller villages, the Regiment fought to break up concentrations of the Insurrectos when they could be found, fought off attacks on its posts and road movements, and engaged in "search and destroy" missions to find and fight an elusive enemy. There were few prisoners taken by either side, (a popular song of the day ran "Beneath the starry flag, we'll civilize them with a Krag") and many Old Guardsmen earned a grudging respect for the Filipino and his weapon and tool of choice—the bolo. After three years, the administration, Army, and the Regiment won the battle for the "hearts and minds" of the

Right: The 3d Infantry arriving in Manila. Notice the transports off the coast. Courtesy of the Library of Congress.

people in the area, and the Philippines remained firmly in American control for fifty years. Malolos, San Isidro, and two consecutive years 1899 and 1900 on Luzon were the campaigns and battles in the Philippines.

The Regiment returned to the United States in April, 1902, to be stationed in Kentucky and Ohio. While the Regiment had been engaged in the war in Cuba and the Philippines, gold had been found in the Alaska Territory, and the influx of people, as well as some concern for the defense of the area because of the Russo-Japanese War, resulted in the Regiment being moved to widely dispersed posts in central and southern Alaska. Operating independently in all weather, companies of the Regiment cleared land and strung wire for the military telegraph which provided communications for these areas. It was unremitting, thankless work, requiring continuous maintenance of miles of wire across land of great beauty and danger. The Regiment was also called to assist in keeping order during natural disasters and in areas where there were no civil authorities. In September, 1906, the Regiment left the territory, taking up station at Forts Lawton and Wright in western Washington State, where, for the first time in two years, men were re-united with their families.

In 1909, the 3d Infantry returned to the Philippines, to the Island of Mindanao in the extreme south of the archipelago. The inhabitants of the island had rebelled against the government in Manila several times in previous years over a series of ethnic, cultural, and religion-based disputes. Most people of the island are Moslems, tied culturally to the people of Malaya, with whom they share their religion, weapons, and political and social customs. A significant problem was piracy, forcing the Regiment to learn small unit amphibious operations. Another three year campaign ensued with the Regiment headquartered at Zamboanga, the provincial capital. The three battalions of the Old Guard found themselves again in an environment of more or less continuous patrolling and small unit operations, punctuated by civil actions, public works construction, and quarantine duty. Combat at the town of Jolo in 1911 earned a campaign credit, but the operations, often in jungle without roads or trails and called "the most difficult terrain in the Philippines" were unremitting over the

whole period. The Regiment returned from the Philippines in 1912.

Keeping the Peace

After three years in the more civilized area of New York State—and reunion with families not seen for three years—the 3d was ordered to the Mexican border. War in Europe and an unstable political situation in Mexico were the reasons, spurred on by an attack on the town of Columbus, New Mexico, by the irregular troops of Poncho Villa. This led to a U.S. action in Mexico led by General John Pershing that came to be called the "Punitive Expedition." Some troops, including the 3d Infantry, remained on the border. When the United States decided to close the unprotected border with Mexico, a large number of troops were kept on the border until the Mexican political situation smoothed out in 1920. A few regiments of regulars, among them the Old Guard, provided a disciplined backbone and cadre of men who trained citizen soldiers for warfare in Europe. Levied several times for men to fill up infantry units ordered overseas, by the end of the war and the occupation duty on the Rhine in 1921, almost every man in the Regiment in 1916 had been ordered overseas. In the Regiment at this time—and destined to rotate through and serve in Europe—were several young officers who were later to become famous, including James A. Van Fleet, later Field Commander in Korea; Henry H. Arnold, Army Air Forces Chief of Staff during WWII, and Matthew B. Ridgeway, Chief of Staff of the Army during the 1950's.

At the end of the border mission, the Regiment was ordered for a short time to Camp Sherman, Ohio, and then to Camp Perry, Ohio. The national rifle matches had been held there for many years, and in 1921, as in previous years, the Old Guard provided a cadre to garrison the post and conduct the matches. In August, 1921, the Regiment was ordered to Fort Snelling, their old home, not seen for twenty years. In the post-war reductions of the army's budget, all funds were limited, and no transportation was available to move the Regiment. Undaunted, they marched.

Leaving Camp Perry in August, 1921, the Regiment marched by road to Fort Sheridan, Illinois, and then to Fort Snelling in November. The snows started early, and the 940 mile exercise was not without incident.

The 3d Infantry training in New Foundland, 1939. This was the first troops sent overseas in anticipation for war with Germany.

The Regiment became acclimatized in this way to its old home in Minnesota and its old mission, soon to be taken up when the regimental commander found that cold weather equipment, including skis, snowshoes, and sled mounts for machine guns had been turned in by the units recently stationed in North Russia. The Regiment settled into a twenty year peacetime mission of training National Guard and Reserve troops from the surrounding states and in the local camp of the Civilian Military Training Corps, where for a few weeks a year, men began to get some military training and a basic introduction to military life. The program was particularly important during the Great Depression, as it provided a structured life, with some pay, for men not otherwise employed.

Hard Times and World War

By the late 1920's, the Army fell on hard times. Having just fought—and, Americans thought, won—the Great War, a wave of anti-military feeling swept over the nation. Disarmament, a wrenching reduction in force for all services, and the drastic shrinkage of budgets for the armed services brought poor morale, unit cohesion problems, and a decline in re-enlistments. The Army sought ways to solve the problems within the constraints imposed.

One way was to use the unique cold weather mission of the Old Guard as an inducement. Sports became more important, and men were sought who could ski, and had cold weather survival skills. Another inducement was an increased Army-wide emphasis on lineage and honors, unit history, and heraldry. The creation of Divisional patches in Europe during WWI—not a new idea in the American service, but unused since the Civil War—had been successful, and the inevitable result was the demand for distinctive insignia for units at lower levels in the Army. In 1922, the Adjutant General requested and received from all regiments in the Army, the units' requests for a distinctive unit insignia for each regiment.That for the 3d Infantry was based on a story that during the earliest days of the Regiment, the men had taken strips of buff colored rawhide and woven them into their black knapsack straps as a distinctive sign. The story is probably apocryphal, but was accepted and is the basis for the adoption of the Buff Strap as a regimental distinction. The Revolutionary War cocked hat insignia, originally used to fasten the strap to the epaulet at the top of the shoulder, was not authorized (the

Left: The 3d Infantry marching from Camp Perry to Ft. Sheridan. They continued the march to Ft. Snelling. This was the longest march in US Military history.

Adjutant General was silent as to how the "knapsack strap" was to be fastened) and was actually forbidden to be worn at least once. The insignia, which came to be called a cockade, was in use again by WWII, and continues in use today.

What became the Second World War began in Europe in September 1939. In the United States, the administration and the armed services come to believe that if we did not help our allies in Europe, we would eventually have to fight the fascists by ourselves. In 1940, at the time of the London Blitz and the beginning of the German U-Boat war, the United States began to negotiate a trade. In return for old, First World War vintage destroyers, the British gave the U.S. use of sites for overseas bases from which troops began a forward defense. Because a declared state of war did not exist between the United States and any other country, the President had limited options in his decision over how to garrison the new bases. Limited to the regular army, he chose the best trained men for cold weather operations for the northern bases. The 3d Battalion, 3d Infantry left Fort Snelling for Newfoundland, becoming the first American unit shipped out of the country in anticipation of our defense of the country against Germany and our entrance into the Second World War. The unit's post was Fort Pepperell, one of the northernmost outposts in the North American security cordon.

While the post was being built in the summer of 1941, the Regiment lived aboard an Army transport, the USAT Edmund B. Alexander, named after the Mexican War commander of the Regiment. The rest of the Regiment moved to Newfoundland eventually and one battalion even moved to Greenland for a time. The unit's mission was to protect the port of St. John's, and help defend the area against raids, all part of a continental defense perimeter set up as a precaution against enemy invasion or enemy intelligence gathering. The area was on the North Atlantic convoy routes to England and Russia, and Newfoundland was used to ferry aircraft from American factories to the front in Europe. The duty was cold, wet, thankless, and absolutely necessary.

By 1943, the menace of the German U-Boats had begun to diminish, and the 3d, after almost three years overseas, rotated to the United States. Stationed in Georgia at Fort Benning and at Camp

Two views of the POW Camp at Weise.

Funeral procession for the unknown soldiers of World War II and Korea.

Butner in North Carolina, the unit reverted to one of its traditional roles, that of training other troops. For a time, the 3d Infantry became the demonstration troops at the Infantry School, teaching men and officers by example in the course of instruction. In the winter of 1944, the course of the war in Europe changed the Regiment's duty abruptly.

The German offensive in the Ardennes resulted in The Old Guard finally being sent to France, arriving in March 1945. Assigned to be a replacement regiment for the 106th Infantry Division, the Regiment began to learn first-hand about war in Europe in the lines around the besieged, German held Channel ports of St. Nazaire and Lorient. An attempted German breakout in the area near Rennes put The Old Guard on alert on 16 April in a usually quiet sector. The orders to move never came, but the 3d earned its campaign participation credit for Northern France. From Rennes, the 3d moved to Germany, near Mainz.

The war was not over, but the great increase in the numbers of prisoners of war was seriously slowing the advance of the American armies. The 106th Division was chosen to handle the influx of prisoners. The Old Guard was given responsibility for camps at Buderich, Limburg, Budesheim, Ditersheim, Mainz, Siershan, Lagstadt, Hergeshausen, Babenhausen, Rheydt, Darmstadt, and others. Literally hundreds of thousands of German prisoner and Russian displaced persons were fed, clothed, medicated, interrogated, classified, and released after April 1945. In the following year, the 3d also assumed responsibility for all Civilian Internee camps in the 7th Army area, a staggering 19 installations, holding 50,000 people, including hospitals, Friendly Witness Barracks, and SS compounds. In all, the Old Guard administered, fed, clothed, and guarded or protected 1,050,000 people in camps scattered over a large part of Germany for a year.

On the 9th of April, 1946, the 81st anniversary of the Old Guard's witnessing the surrender at Appomattox, the unit moved to Berlin to become part of the occupying forces of that capital. Staying until November, the unit was relieved by the 16th Infantry and was, for the first time, inactivated. Caught in the post-war demobilization, The Old Guard's colors and records were sent back to the United States on 20 November 1946.

ABOVE: Tomb of the Unknown Soldier, 1950's.

BELOW: General George S. Patton's funeral in 1945. The Old Guard oversees and maintains all government funeral processions for the United States.

Presentation of Colors during the Reactivation Ceremony in 1948.

Vietnam & the Cold War

On 6 April 1948, the 3d Infantry Regiment was reactivated on the Capital Plaza in Washington, DC. What came to be known as the "Cold War" created a need for greater protection of the capital, national leaders, and public property. It was decided that The Old Guard, as the oldest Infantry regiment, would be the ideal choice. The mission to protect the capital had been performed by Military Policemen of the 703d and the 712th MP Battalions since the end of the war. The men of the 703d were transferred to the lst Battalion, 3d Infantry, and those in the 712th became the 2nd Battalion, 3d Infantry, with a large number of recruit trainees from Fort Dix, New Jersey added to both battalions. The ceremonial mission in MDW had been performed (since 1943) by the Military District of Washington Ceremonial Detachment. The Detachment became part of Company A of the Old Guard in 1950, beginning the 3d's ceremonial mission in the capital. With the Ceremonial Detachment came the mission to perform burials in Arlington Cemetery, diplomatic arrivals and departures, the perpetual guard at the Tomb, as well as caissons and horses.

The Old Guard had an interest in its history for most of the 19th century, and has maintained a regimental collection of artifacts, uniforms, flags, and documents since before the Civil War. Because of WWII it almost lost them. When the last elements of the Regiment left Fort Snelling in 1942, this collection was put for safekeeping at the Minnesota Historical Society at St. Paul. With the reactivation of the Regiment in 1948, there were no members of the unit with any knowledge of the past history of the Regiment, or the regimental collection. Until 1950 or 1951, the Regiment's physical ties with its past remained lost. A single exception was a photograph found of the Chapultepec Baton, and a replica (not to scale) made with materials from a White House renovation, presented to the Regiment by President Truman. The (apparently) chance assignment of MSG Jack Watts to the Regiment at Fort Myer saved the collection. Watts, who had soldiered in the 3d before the war, remembered the items exhibited at Fort Snelling, and asked the commander, COL William W. Jenna, where they were. Jenna sent Watts to Fort Snelling to look for the collection, which was eventually located and returned to the Old Guard. This collection, one of the few regimental collections in the Army, formed the nucleus of the Old Guard Museum when it was begun in 1957.

For a time during the Korean War, the Old Guard was severely hampered by the constant rotation of infantry troops from the United States to units based in Korea. Company H, which was then a provisional company of basic trainees being groomed for the ceremonial duties of the Old Guard, was shipped out of Fort Myer almost immediately after graduation in 1951. These men served as replacements in the 8th Army and did not return to their Old Guard duties.

During this period came the development of the changing of the guard at the Tomb from a simple guard mount to the formalized, ritualized ceremony of a later day, and the integration of the Army and the Old Guard. The first black soldier became an Old Guardsman in 1953; the first became a Tomb Guard in 1961.

Since 1948, the Regiment has been reorganized several times, becoming Battle Groups, a series of autonomous battalions, and reduced eventually to a single reinforced battalion. In one of the Army-wide reorganizations, in 1957, the functional regimental system under which the Old Guard had been created in 1784 ended. Thereafter, there would be no separate regimental headquarters, but only unrelated units having the same lineage and honors. In 1963, the 2d Battle Group, 3d Infantry, was inactivated in Korea, having been created as the 2d Battalion at Fort McNair in 1948. With the coming of the war in Vietnam, a

Chapultepec Baton and Swagger Baton, 1921.

new 2d, and a 4th, 5th, 6th, and 7th Battalions were created. All of these were autonomous, having a separate command and staff from the 1st Battalion at Fort Myer, and never serving under a single regimental headquarters. The 3d Battalion, in recent times a USAR unit in the Regiment's old home at Fort Snelling, was activated in 1959 and inactivated in August 1994.

In 1966, the 2d and 4th Battalions were activated, the 2d at Fort Benning and the latter at Schofield Barracks, Hawaii. The 2d Battalion was assigned to the 199th Infantry Brigade, eventually serving in the area around Saigon (now Ho Chi Minh City), and the 4th Battalion served in the 11th Infantry Brigade and then the 198th Infantry Brigade of the Americal Division in the I Corps area near the northern border of South Vietnam. Both battalions were inactivated in 1970. Neither battalion was very much aware of the other's existence in Vietnam, but both, at various times, requested the assistance of the 1st Battalion at Fort Myer in providing histories of the Old Guard and other information. In their four year's existence, both battalions were decorated, the 2d receiving a Valorous Unit Award and the 4th's Company E Reconnaissance Platoon receiving a Presidential Unit Citation. The 5th, 6th, and 7th Battalions were training battalions of the 6th Infantry Division constituted at Fort Campbell, Kentucky, in 1967 and were inactivated there in 1969.

ABOVE: Eisenhower's Inaugural Ceremony, 1957.

BELOW: The Fife and Drum Corps of The Old Guard, 1960.

The 3d Infantry
"Old Guard"
serving in Vietnam.

Then & Now

In every conflict (and most larger operations) since Vietnam, The Old Guard has been levied for soldiers to serve in units overseas, often troops with a specialty needed desperately during a deployment. These soldiers served not as Old Guardsmen, (since the unit is not deployed) but with other units.

The requirement for the Old Guard to represent all of the Army, regardless of gender, has created special circumstances and solutions in the 3d Infantry in the past, and will again. For the first 194 years of its history, no women served in the ranks of the Old Guard. Because of its light infantry mission, there are still some jobs in the Old Guard from which women have always been excluded by law. But the unit has the responsibility to represent all of the Army in its ceremonies. For most of the past twenty years, The Old Guard has had women serving in its ranks for ceremonial purposes, and as musicians. Beginning in 1978, the unit recruited women to perform ceremonies with a line company, creating a Female Detachment in E Company. The Old Guard Fife and Drum Corps recruited its first female musician in 1982. For a few years at the end of the 1980's, there were again no women marching with the line companies. A new, more permanent solution was found when the 289th MP Company (-) was attached to the 3d Infantry on November 1st, 1994. This action was based on the precedent of the Military Police Battalions that were re-designated to re-activate the 3d Infantry in 1948. The 289th MP Company is a decorated unit with service during WWII and Korea. The soldiers of the 289th of both genders are fully integrated into the Army's oldest active infantry regiment, the Old Guard.

The 1st Battalion at Fort Myer, Virginia is a unique unit, there being no other in the world like it. Although most nations have

units in their armed services that perform ceremonial duties, none has the mix of missions peculiar to The Old Guard. Its parts have grown and developed over the last fifty years to provide capabilities unknown elsewhere. It is the only US Army unit to have official historic uniforms; the only to have its own field music; and the only to have and use horses in performance of its official duties. Since 1950, only Old Guard soldiers have guarded the Tomb of the Unknowns. While it is not the only service to support government operations in the capital, it is the lead element of the leading service.

Known by its regimental designation of 3d US Infantry (The Old Guard), the performance of its ceremonial mission, the fulfillment of its obligation to present the best impression of the United States Army to the world. Its observance of Old Guard customs and standards, are all now a part of a heritage of more than two hundred years of service. Such a mission would be honorable duty for any unit. It is a fitting mission for The Old Guard.

OPPOSITE: *From left: George Osborne, Colonial Uniform, Flag Bearer; Joseph M. Lovell, Tropical Worsted; Frankie Solace, Olive Drab; John E. Hummell, Blue. Courtesy of Frankie Solace.*

BELOW: *The US Army Caisson Platoon, bearing the body of General of the Army Omar Bradley, approaching Arlington National Cemetery, April 1981. Courtesy of Dean R. Lindeman.*

OPPOSITE: Honoring the Veterans, an American flag being presented to a service person's family.

HONOR GUARD

76
BLACK JACK

Time Line of Events

The dates of origin for 3d Infantry and some of its elements are below:

3 June 1784
The Regiment, called the 1st American Regiment

May-October 1815
Redesignation as the 3d US Infantry

1922
First use of historical uniforms for official purposes (in a color guard)

20 November 1946
Inactivation

6 April 1948
Reactivation

1950
MDW Ceremonial Company becomes A Company

10 April 1952
President Truman presents the Presidential Baton to the Regiment, and names it "Honor Guard to the President"

1957
Organization of the US Army Drill Team

1960
Organization of the Fife and Drum Corps

1963
Official "traditional designation" of 3d Infantry as "The Old Guard" recognized

10 September 1971
H Company was activated

December 1973
Activation of Commander-in-Chief's Guard

25 August 1994
Inactivation of the 3d Battalion, 3d Infantry

1 November 1994
289th Military Police Company attached to the Old Guard

1 November 1995
Fife and Drum Corps becomes a separate company

11 September 2001
Unit Activated for contingencies involving rescue and recovery for the disaster site at the Pentagon

A Company, 1st Armed Forces Day Parade, Washington DC, May 1950.

US Army Unit Awards

1st Battalion: Meritorious Unit Commendation
awarded for the period which includes the Nixon inaugural and the Johnson funeral (20 Jan 1969—25 Jan 1973), awarded 17 May 1974.

1st Battalion: Army Superior Unit Award
awarded for the period which includes 1 May 1984-1 February 1985, internment of Vietnam Unknown awarded 27 October 1985.

1st Battalion: Army Superior Unit Award
awarded for the period which includes the Presidential Inaugural activities 5 January 1993-10 February 1993, awarded 1 September 1993.

2d Battalion: Valorous Unit Award
streamer embroidered Saigon-Long Binh, 1968

4d Battalion: (Reconnaissance Platoon, E Company) Presidential Unit Citation 1968.

Ceremonial Co. I.G. inspection on Drill Field, 1949. Courtesy of Frankie Solace.

Campaign Participation

3d Infantry Regiment

During the Civil War, it became common for units to paint the names of the battles in which they had been participants on their colors, usually in the red stripes of the national standard. In 1890, when the size of colors carried in the field was reduced, the Army standardized honors by directing that they be engraved on a silver band around the flagstaff. In 1919, the current system of streamers was adopted.

In the order in which earned:

Miami
Canada
Chippewa
Lundy's Lane
Seminoles
Palo Alto
Resaca de la Palma
Monterey
Vera Cruz
Cerro Gordo
Contreras
Churubusco
Chapultepec
Comanches
New Mexico 1856
New Mexico 1857
New Mexico 1858
New Mexico 1860
Texas 1861
Florida 1861
Bull Run
Florida 1862
Peninsula
Manassas
Antietam
Fredericksburg
Chancellorsville
Gettysburg
Virginia 1863
Appomattox
Montana 1887
Santiago
Malolos
San Isidro
Luzon 1899
Luzon 1900
Jolo 1911
American Theater without Inscription
Northern France
(Vietnam) Counteroffensive Phase II
(Vietnam) Counteroffensive Phase III
(Vietnam) Tet Counteroffensive
(Vietnam) Counteroffensive Phase IV
(Vietnam) Counteroffensive Phase V
(Vietnam) Counteroffensive PhaseVI
(Vietnam) Tet 1969 Counteroffensive
(Vietnam) Summer-Fall 1969
(Vietnam) Winter-Spring 1970
(Vietnam) Sanctuary Counteroffensive
(Vietnam) Counteroffensive Phase VII
(Vietnam) Consolidation I

289th Military Police Company (minus)

(attached 1 November 1994)

European-African-Middle Eastern Theater without inscription
Korea CCF Intervention
Korea First UN Counteroffensive
Korea CCF Spring Offensive
Korea UN Summer-Fall Offensive
Second Korean Winter
Korea, Summer-Fall 1952
Third Korean Winter
Korea, Summer 1953

289th Unit Awards

Meritorious Unit Citation - Korea
Meritorious Unit Citation - Korea 1953
Meritorious Unit Citation - Korea 1954

Opposite: 3d Platoon, A.P. Hill, October 1967. Courtesy of J.P. Ganio.

Below: Ceremonial Co. returning from Retreat Parade, Ft. Myer, Virginia, 1949. Lt. Haus was Officer in Charge.

Medal of Honor Recipients

The Medal of Honor is the highest award for valor in action against an enemy force which can be bestowed upon an individual serving in the Armed Services of the United States. Generally presented to its recipient by the President of the United States of America in the name of Congress, it is often called the *Congressional Medal of Honor*.

There have been four Medal Of Honor recipients in the history of the 3d Infantry, one of those being awarded posthumously.

Sergeant James Fegan

Company H, 3d US Infantry

Place and date:
Plum Creek, Kansas, March 1868

Citation:
While in charge of a powder train enroute from Fort Harker to Fort Dodge, Kansas, was attacked by a party of desperadoes, who attempted to rescue a deserter in his charge and to fire the train. Sergeant Fegan, single handed, repelled the attacking party, wounding two of them, and brought his train through in safety.

Corporal Leander Herron

Company A, 3d US Infantry

Place and date:
Near Fort Dodge, Kansas, 2 September 1868

Citation:
While detailed as mail courier from the fort, voluntarily went to the assistance of a party of four enlisted men, who were attacked by about 50 Indians at some distance from the fort and remained with them until the party was relieved.

Private Oscar Burkard

Hospital Corps, United States Army
Serving with the 3d Infantry

Place and date:
Leech Lake, Minnesota, 5 October 1898

Citation:
Bravery in action against hostile Indians. PVT Burkard rescued and tended to several wounded in the action, continuously exposing himself to sniper fire. His award was the last given for actions against hostile indians.

Corporal Michael Fleming Folland

Company D, 2d Battalion, 3d Infantry, 199th Infantry Brigade

Place and date:
Long Kahn, Province, Republic of Vietnam, 3 July 1969

Citation:
For conspicuous gallantry and intrepidity in action at the risk of his life above and beyond the call of duty. Corporal Michael F. Folland distinguished himself on 3 July 1969 while serving as an ammunition bearer with the weapons platoon of +Company D, 2d Battalion, 3d Infantry, 199th Infantry Brigade during a reconnaissance patrol mission in Long Kahn Province, Republic of Vietnam. As the patrol was moving through a dense jungle area, it was caught in an intense crossfire from heavily fortified and concealed enemy ambush positions. As the patrol reacted to neutralize the ambush, it became evident that the heavy weapons could not be used in the cramped fighting area. Corporal Folland dropped his recoilless rifle ammunition, and ran forward to join his commander in an assault on the enemy bunkers. The assaulting force moved forward until it was pinned down directly in front of the heavily fortified bunkers by machinegun fire. Corporal Folland stood up to draw enemy fire on himself and to place suppressive fire on the enemy positions while his commander attempted to destroy the machinegun positions with grenades. Before the officer could throw a grenade, an enemy grenade landed in the position. Corporal Folland alerted his comrades and his commander hurled the grenade from the position. When a second enemy grenade landed in the position, Corporal Folland again shouted a warning to his fellow soldiers. Seeing that no one could reach the grenade and realizing that it was about to explode, Corporal Folland, with complete disregard for his own safety, threw himself on the grenade. By his dauntless courage, Corporal Folland saved the lives of his comrades although he was mortally wounded by the explosion. Corporal Folland's extraordinary heroism, at the cost of his life, was in keeping with the highest traditions of the military service and reflects great credit upon himself, his unit, and the United States Army.

3d U.S. Infantry Commanders

First Regiment - 1784-1815

1784-1790
Brigadier General Josiah Harmar
1790-1803
Colonel John F. Hamtramck
1803-1808
Colonel Thomas Hunt
1808-1815
Colonel Jacob Kingsbury

Third Regiment

1815-1818
Colonel John Miller
1818-1821
Colonel Joseph Lee Smith
1821-1825
Colonel Ninian Pinkney
1825-1834
Colonel Henry Leavenworth
1834-1840
Colonel James B. Many
1840-1851
Lieutenant Colonel Ethan Allen Hitchcock
1851-1855
Colonel Thomas Staniford
1855-1861
Colonel Benjamin L. E. Bonneville
1861-1862
Colonel Charles F. Smith
1862-1869
Colonel William Hoffman
1869-1871
Colonel George Washington Getty
1871-1879
Colonel DeLancey Floyd-Jones
1879
Colonel Luther Prentice Bradley
1879-1888
Colonel John R. Brooke
1888-1895
Colonel Edwin C. Mason
1895-1903
Colonel John Henry Page
1903-1904
Colonel H. L. Haskell
1904-1910
Colonel T. C. Woodbury
1910-1914
Colonel Henry Kirby
1914-1915
Colonel William A. Mann
1915-1917
Colonel Robert L. Hirst
1917-1918
Colonel Frederick R. Day
1918-1919
Colonel Gustave A. Weiser
1919-1921
Colonel Paul Giddings
1921-1923
Colonel Alfred W. Bjornstad
1923-1924
Colonel Elliot Malloy Norton

1924-1925
Colonel William Roderick Sample
1925-1928
Colonel William E. Walsh
1928-1931
Colonel Walter C. Sweeney
1931-1932
Colonel David L. Stone
1932-1934
Colonel Charles J. Nelson
1934-1936
Colonel John Randolph
1936-1937
Colonel Charles F. Thompson
1937-1939
Colonel Martin C. Shallenberger
1939-1940
Colonel Maurice D. Welty
1940-1943
Colonel Paul H. Brown
1943-1944
Colonel Paul N. Starlings
1944-1945
Colonel John S. Moore
1945-1946
Colonel Herbert Vander Heide
1948-1950
Colonel James V. Cole
1950-1952
Colonel William W. Jenna
1952-1954
Colonel Dennis M. Moore
1954-1955
Colonel Robert L. Cook
1955-1956
Colonel Fielder P. Greer
1956-1958
Colonel Mark F. Brennan
(1st Battalion, 3d US Infantry)
1958-1960
Colonel Christopher W. Chaney
1960-1962
Colonel Richard M. Lee
1962-1963
Colonel John W. Gorn
1963-1964
Colonel Richard E. Cross
1964-1968
Colonel Joseph B. Conmy, Jr.
1968-1970
Colonel Robert M. Daugherty
1970-1972
Colonel Myron E. Lee, Jr.
1972-1974
Colonel Harvey H. Perritt, Jr.
1974-1976
Colonel Robert H. Clark
1976-1977
Colonel Robert D. Wiegand
1977-1979
Colonel Donald C. Hilbert
1979-1981
Colonel Stanley G. Bonta
1981-1982
Colonel Don Phillips
1982-1985
Colonel Robert G. Krause
1985-1987
Colonel William R. Williamson
1987-1989
Colonel James F. Hennessee
1989-1991
Colonel Barrie E. Zais
1991-1993
Colonel James F. Hunt
1993-1995
Colonel Stephen A. Nash
1995-1997
Colonel David H. Huntoon, Jr.
1997-1999
Colonel Gregory C. Gardner
1999-2001
Colonel Thomas M. Jordan
2001-2003
Colonel James F. Laufenburg
2003-2005
Colonel Charles L. Taylor

Opposite: *Black Jack and member of the Honor Guard at Arlington National Cemetery during the winter of 1976.*

Below Top: *Saluting the Tomb of the Unknown Soldier.*

Below Bottom: *Burial flags draped across caskets before a military funeral.*

Right: *Member of the Old Guard presiding over the Tomb of the Unknown Soldier.*

Below: *The Old Guard presiding over a military official's funeral in Washington D.C.*

Opposite Top: *The Old Guard practicing for a funeral precession.*

Opposite Bottom: *On March 4, 1921 Congress approved the burial of an unidentified American soldier from WWI in the plaza of the new Memorial Amphitheater. This became known as the Tomb of the Unknown Soldier.*

HERE RESTS IN
HONORED GLORY
AN AMERICAN
SOLDIER

Opposite Bottom: Members of the 3d Infantry jogging through Arlington's grounds.

Left: Bravo Company, training for the Global War on Terror at the Horn of Africa.

RIGHT: *NCOIC, US Army Drill Team.*

OPPOSITE: *In formation near the Capitol, the Old Guard Rifle Team prepares for President Ronald Reagan's Funeral.*

BOTTOM: *Presenting the State and Territorial Colors.*

Bottom: President George W. Bush assisting with the wreath at President Ronald Reagan's funeral.

Opposite: Funeral procession, President Ronald Reagan.

Below: Standing guard over the Tomb of the Unknown Soldier.

HONOR GUARD

Below: "Flags In," an annual ritual in Arlington National Cemetery.

Opposite: Caparison horse with President Reagan's personal brown riding boots.

Below Top: Bivouac of the dead, Arlington National Cemetery.

Below Bottom: Edinbugh Military Tattoo, Castle Esplanade.

BELOW TOP: The Old Guard procession during a military funeral at Arlington.

BELOW BOTTOM LEFT: "Flags In" at Arlington National Cemetery.

Moments to Remember

PLT Sgt. Wamer aboard Conversano Beja on duty, Ft. Myer, Virginia. Courtesy of Michael D. Juers.

Memorable Moment From TOG

by Silvio R. Annunziata

JFK had been assassinated, the sadness was everywhere. On Sunday, Nov. 24, 1963, I was assigned to a street cordon right at the intersection of 16th and Pennsylvania. There were thousands of people there; the only sound that could be heard was the sobbing of people in the crowd. Then the drums started, growing louder and louder, windows were vibrating in the surrounding buildings. I looked up to see a man violently yanked from a second story window across the street from me. The procession started down 16th Street, when it reached my position, I remember the almost involuntary head nod and then the rifle salute in tandem with another Old Guard soldier across the street from me. As the cortege passed me, the crowd surged forward and filled in Pennsylvania Avenue and followed the procession as it made its way toward the Capitol.

On Monday, November 25th, after an initial positioning on 4th Street, we moved to the Lincoln Memorial. Again the silence was deafening, Red Cross workers would stop occasionally to see if we were all right, we stood at parade rest for (I believe) four hours. Then the final procession started going by, every world leader walked passed me, Churchill, DeGaulle, Haile Selassie and others I can't remember. There were miles of limousines. The Marine Band marched passed and I was struck in the head by the tuba as they extended the rank to compensate for the traffic circle, the silence

Cortege for President John F. Kennedy, 1963.

was unbelievable, I kept thinking, "How could so many people be so quiet, except for the sobbing." Then the missing man formation flew over, the 21-gun salute, the sound of taps, so loud it seemed like it was coming from right in front of me.

As the sound of taps faded, the only sound I could hear was a few leaves rustling softly down the street against the curb. To this day, every time I hear the sound of the leaves moving along the ground, I am instantly back there, standing at the Lincoln Memorial.

Charlie Guard

Written by SFC James R. Aux

The best of The Old Guard
Who is the Charlie Guard soldier?
He is the proud bearer of our nation's honored dead.
He is the best in a 21 gun salute.
He is the tallest and sharpest on Summerall Field.
He is the unsung RTO, supply specialist and armorer.
He is an infantryman who knows that few, but his fellow Soldiers, realize his worth.
He is a flag bearer at the Jefferson Memorial, an usher at Constitution Hall, a member of the color team at the Tomb and the "present man" on a full honor funeral.
He is a CQ runner, a guidon bearer in "The Old Guard Run" and a "Top Kick" who commands, "FALL IN!"
But, most of all...
He is the soldier whose heels click the loudest when he hears,
"Old Guard, ATTENTION."

A Very Emotional Event

by Lt. Col. Albert G. Bertaccini

The story of the 3rd Infantry Regiment when it joined other regiments and infantry units to re-group the 106th Infantry Division at Rennes, France in 1945.

Above: *Tomb Guard in the snow, 1980.*

Below: *World War II, machine gun training.*

The 106th Infantry Division has been almost annihilated at the Battle of the Bulge and not to have the Germans realize that a complete infantry division has been knocked out – it was decided to re-group a new-old infantry division at a parade field in Rennes, France.

Flags of the original units of the 106th Infantry Division were carried by the remnant soldiers (400+) and marched to the center of the parade field. There they met the complete new division and swapped flags with the newly formed 106th Infantry Division, the "Screaming Lions"!

Then in parade form, the new division marched proudly in review, past the remnants of the old 106th Infantry Division.

The new 106th Infantry Division went on to support Gen. Patton's 3rd Army in the attack of the Germans at the Rhine River and on into Austria and Czechoslovakia!

Rifle Team in Arlington National Cemetery.

A Huge Success

by John C. Boyle Jr.

Arlington National Cemetery, October 17, 1998: almost 600 arborists from 20 states and 80 companies brought their chain saws, chippers and bucket trucks to Arlington National Cemetery to provide tree care to many of the 14,000 trees at the Cemetery as part of National Arborist Association's (NAA) "National Day of Service."

The National Day of Service is part of an ongoing volunteer effort by the association to improve the environment by providing professional tree care to our nation's public and private green spaces. NAA has sponsored similar events at Ellis Island, Liberty Island, and Independence Hall, as well as at Arlington Cemetery in 1993.

Working at 25 different zones throughout the 612-acre property, including areas near the Women in Military Service in America Memorial, the grave of President John F. Kennedy, and the Tomb of the Unknown

Soldier, the arborists climbed trees ranging in height from 20 feet to over 80 feet, provided fertilization, cabling, pruning, and other preservation and maintenance work. Representatives from NAA explained the importance of tree care to visitors at the Cemetery.

General Walton H. Walker is placed on the caisson for removal to the gravesite in Arlington National Cemetery. General Walker was killed as the result of a jeep accident while Commanding the Eighth Army in Korea. Reverend John Walthour conducted the service. Cpr. Gerald Comer is holding the black horse with the turned boots. Courtesy of Gerald Comer.

"Our members came here ready to work, and accomplished more real tree work in seven hours than the Cemetery will be able to accomplish in the next two or three years," explained Paul Wolfe, NAA President. "This National Shrine is symbolic to every American who has ever served in the armed forces, or who knows anyone who has served. The dollar value of the gift is approximately $400,000. The benefit to the environment is truly invaluable. The NAA is pleased to have provided this gift to the nation by contributing to the importance of tree care in protecting the aesthetic and economic value of land—whether it be this critical piece of our country's landscape, or a backyard."

Jack Metzler, Superintendent of Arlington National Cemetery added: "today's National Arborist Association donation will enhance the appearance of our trees and landscaping far beyond our own capabilities. This is a great contribution to this property, our nation, and millions of visitors who tour these grounds."

NAA is a nonprofit organization dedicated to the advancement of commercial tree care businesses. Since 1938, NAA has been committed to educating the public about the environmental benefits of professional tree care through public service events such as National Day of Service.

Guard Of Honor

by Ralph Casado

Just across from the Capitol, on the Virginia side of the Potomac, lies in solemn beauty, the site of Arlington National Cemetery. Visitors will recall the measured rows of plain white markers and uncluttered grounds. Side by side lie Continental troopers, Yankees, Johnny-rebs and G.I. Joes.

I will always recall that first week I was posted there. It was summertime. August days in New York were hot, but here the heat-dulled air and nagging buzzing of summer pests made my well-fitted uniform a little more unbearable; as rigidly at attention as I stood I could still slip my tongue, unnoticed, on my lips to lap in the rivulets of moisture trickling from under the sweatband of my military cap.

I was beginning to dread the routine that the next year held in store, as the excitement of taking part in a military funeral was wearing off. With all its quiet pageantry, the regulated, almost staged scene that I had witnessed and been part of several times a day for the past five days with complete disinterest, suddenly became very real. The seemingly, until then, faceless mourners became very human. The old general's widow took on a personal character, as she limply accepted the tri-cornered neatly folded flag, "a token from the Government of The United States for faithful and meritorious service, etc..."

With the echo of the three volleys we had just fired still ringing against the invisible roof of the sky, I became choked up. In a seemingly unimportant moment my unconcerned youthful attitude towards death was shattered, and time seemed to become very precious. At that moment I felt as if the very seconds, ticking away, were melting my life.

Thinking back to that time does not give me discomfort now. It seems as if that instant was one of my more shocking steps to maturity.

Following is a letter sent to my father:

Headquarters
Third Infantry Regiment
Fort Myer, Virginia

19 August 1955

Mr. Ralph Casado, Sr.
6423 18th Avenue
Brooklyn, New York

Dear Mr. Casado:

Your son, Ralph, has recently joined the 3d United States Infantry Regiment with Headquarters at Fort Myer, Virginia. The Fort is located just across the Potomac River from the Nation's Capitol. The present duties of the Regiment include furnishing Honor Guards and security details for the Washington area as well as the necessary administrative and functional support of the unit. Your son has been selected to participate in these duties. Although it is impracticable for the Army to make permanent assignments in these troubled times, it is hoped that your son will remain with us during his normal tour of duty.

Personnel of the 3d Infantry are quartered in permanent brick buildings and are provided with well balanced meals, the preparation of which equals the best in the Army. A beautiful Post Chapel is located at Fort Myer where religious services of all faiths are provided and where chaplains are available for consultation and advice.

The Post has two enlisted men's clubs as well as two swimming pools, two service clubs, a large gymnasium and a craft shop. In addition, there is an Information and Education Center where personnel may obtain advice and guidance concerning the furtherance of their education. A number of off-duty classes by the Army, the George Washington University, and the University of Maryland are readily available. The 2nd Battalion of the 3rd Infantry located on the Potomac River at Fort Lesley J. McNair, Washington, DC, offers facilities similar to those of Fort Myer.

The 3d Infantry, or the "Old Guard", as it is popularly known, is the oldest Infantry Regiment in the Regular Army. Its inception dates back to 1784. The selection of your son for assignment to this unit is something of which both you and he may be justly proud. It is hoped that through this letter I have been able to give you some pertinent information concerning Ralph's surrounding.

Sincerely yours,

Fielder P. Greer
Colonel, Infantry
Commanding

My Story

by Dennis W. Choplin

In May 1961, while participating in a demonstration at Andrews Air Force Base on Armed Forces Day the worst happened. While putting on the demonstration of an Infantry Unit taking an objective supported by tanks, all units were on line taking a mock objective all firing blanks including the tanks. I was a driver of one of the tanks commanded by SFC Jones, a very good tank commander (TC). As we were readying to circle the objective, the TC told me to go right. As I started turning, he started turning the main gun back on the objective. We were driving with "open hatch" (for looks only). Drivers were wearing a type helmet not designed for tank drivers (all one piece). As the main gun started turning to the left, I noticed that my head would not clear the front of the turret under the main gun. We knew before the exercise that the intercom between myself and the TC was not working. I could only hear him. As the main gun started my direction, I reached for a lever that would drop the seat. I could not locate it nor could I move the seat in anv direction. The turret hit my helmet, pushing my head against the hatch opening. I could hear the crushing, the helmet popped off my head and crashed to the ground. I reached for the switch to turn off all power but wanted to wait until the turret had cleared my head. Just as it passed, I flipped the switch, cutting all power. I could hear the TC hollering for me to move on. As the tank stopped, I thought any second I would black out and that would be the end of me. Sometime earlier there had been reports of three such incidents where two of the three had caused the death of the driver. I crawled out with blood all over the side of my head and all over my fatigue jacket. As I came into view of the TC, he knew exactly what had happened. I crawled to the side of the turret, he grabbed me and told the loader to turn the tank around and get on the runway and head for the aid station. The loader was driving the tank so fast that when he attempted to stop, each time he hit the brakes, the main gun would go down almost hitting the runway. He was really moving the tank. The demonstration went on as planned. The spectators probably thought it was part of the exercise.

Tank Platoon of HHC 1st BG, 3d Inf.

I was taken to Andrews AFB Hospital for treatment. I had cuts over my right eye that had to be stitched up. The right side of my face was really scraped. I had to wear a black patch over my right eye for a few weeks.

Just a few days later while in the motor pool cleaning the brass from the .30 cal coaxial machine gun from under the turrets, my TC, SFC Jones, was standing on the sponson box, leaning against the turret looking down inside the tank. The tank next to him was also cleaning from under the turret. They would remove a little then reach up and move the turret a little more to get to more of the brass. In doing so, the main gun tube inched closer to hitting the next tank. It finally did, hitting SFC Jones in the back, stopping just in time, pinning him to the side of the turret. Someone saw what happened and alerted the other tank personnel. People were so excited they forgot to hold SFC Jones as they turned the main gun back, and he fell off to the ground. Luckily he survived with only damaged ribs.

I was also assigned to HHC 4th BN, 3rd Inf. (The Old Guard) while in Vietnam. I was located on FSB San Juan Hill in the middle of a free fire zone. While there I sprayed a lot of agent orange around. I used several hundred gallons. Two experiences come to mind: the first was when the Battalion Commander wanted me to take a flame thrower via helicopter with security and try to burn off the top of a mountain for a new fire support base. When the chopper landed, I was in the middle of a free fire zone with a small amount of security, which I could not see or hear as I had no radio. I went to the upwind edge of the mountain ledge to shoot flame over the side so I would burn off the top. As I pulled the trigger, I spun around and fell down. I didn't know I was standing on a loose rock. As I fell landing on my back, I looked up seeing the flame going straight up. I thought "oh no, the flame may come back down on me." Luckily it didn't. I finally let go of the trigger. I proceeded to burn the hill top. As the chopper came in to get me I could not see a thing because the blades were throwing up the black ash from the burned weeds. I was sweating very badly, and the ash stuck to me like glue. I was totally black and had to find the chopper by sound only. I really worried about walking into the rotor blades as we had just lost a BN CO from such type accident. The gunner jumped off and got me.

Another time the S-3 wanted me to put agent orange into prophylactics sitting in No.

Courtesy of Dennis Choplin.

10 cans and go by helicopter and drop them on rice paddies. I took seven cans loaded and started dropping but had a problem with agent orange blowing all over the helicopter and door gunner so I had to stop. What a mess. The door gunner and I were soaked with agent orange.

While assigned to the 1st BN 3rd Inf. (The Old Guard), I had the privilege of having Lt. Humbert R. Versace as my platoon leader. He was later captured in Vietnam and was a prisoner for two years before being executed by the North Vietnamese on 26 September 1965. He was later awarded the Congressional Medal of Honor by President Bush on 8 July 2002. His citation can be read on the internet by typing in "Capt. Humbert R. Versace." He did it by the book. Please read "*Five Years To Freedom*" by James N. Rowe. It's a true story of a Vietnam POW and tells about Captain Versace.

Why We Guard The Tomb

by Neale Cosby

Forty plus years ago, I was the platoon leader of the guards at the Tomb of the Unknown Soldier in Arlington National Cemetery. One day I was standing on the plaza inspecting a guard change when a little old lady asked me the usual set of questions.

Are there guards here at night? Answer: yes, 24 hours a day, 365 days a year. And it has been so since 1936. Every second, of every minute, of every day - rain, shine, snow, sleet, a guard is here day and night.

Then she asked how are the guards selected? I cited the high character

Plt. Sgt. Wapler, Ceremonial Company, 3d Inf., Ft. Myer, Virginia, holding President Truman's hat, 1948.

requirements for duty at the Tomb. In short, I told her they must be good soldiers and good citizens.

Following the guard change she came back with more questions. She was persistent. I can still see her face today. She asked: Do the guards like the duty here? I said, Yes. They feel it is an honor to continuously render, in their own small soldierly way, this nation's highest honors to the Unknown Soldier.

She said what do you mean? I explained that the highest honor that this country renders to anyone is a 21-gun salute. The President gets a 21-gun salute. And then I told her that that this is what this lone sentinel is doing, continuously. Then I pointed to the sentinel on the mat. In a whisper, I counted 21 seconds as he paused at each facing movement — one thousand one, one thousand two and so on. I counted the 21 steps as he crossed the mat from one side of the Tomb to the other, always keeping an eye on the Tomb and keeping his body between the Tomb and the weapon he carries. Thus, rendering the highest honors with his 21-count, continuously.

At the end, she began to cry and she said, "You may wonder why I'm asking these questions, but you see, I have a personal interest here. This is the only place I have to pay my respects to my son who did not return from the war."

With that she departed, leaving me touched for life and forever answering the question of WHY WE GUARD THE TOMB.

Burials I Won't Forget

by Richard L. Craig

I played on the post baseball team from March to September of 1952. After the

Company A, Arlington National Cemetery, 1973. Included in picture from left: Bob Oldham, Gary Epperly, Ed Riley, Ron Kosola. Courtesy of Edward J. Riley.

season was over, I participated in many honorary burials. One burial that I will never forget and brought great sadness to our hearts was Company Commander Captain McCoy's. Captain McCoy had volunteered for a second tour of duty to Korea where he had previously won a battle field commission shortly after Thanksgiving of 1952. Unfortunately, he was killed in action in less than two weeks. We buried him, as best as I can recall, on or about December 18th in Arlington National Cemetery. That burial was a tough one.

Another burial that had a great impact on me and the rest of the new trainees was our Corporal Bayonet Specialist who trained us in bayonet fighting while stationed at Fort Myer during April to July of 1951. We were bivouacked either at Fort Meade or Fort A.P. Hill. This Corporal Bayonet Specialist was then was sent to Korea and shortly thereafter killed. This became the first burial that most of the new trainees and I were assigned to perform. The irony of this story - he was killed by a bayonet.

Treasured Memories

by Roy S. Cummings

After fifteen difficult months in Korea, tour extended because of the Berlin Crisis, I was assigned to Company A, 1st Battle Group, of the 3rd Infantry in September of 1962. We were stationed at Fort Leslie McNair in Washington DC. In contrast to the bleak countryside in Korea, this was like a college campus. My job in the third platoon was to provide a casket team for the burial of the deceased military personnel and their spouses. The majority of the funerals were at Arlington, although we also went to a small cemetery outside DC in Maryland, and also to Culpeper, VA. We enjoyed that trip because we would leave in the early morning and return by five. Master Sgt. Domblinski would be the NCO in charge and he always had us stop for lunch after the ceremony at a little restaurant in Culpeper. After eating all our meals in the mess hall, a lunch in a real restaurant was some treat for us. We also performed cordons for visiting dignitaries from all over the world. We marched around the Washington Monument on Sunday during the summer months. That gave us Wednesday afternoons off. We always coveted those weekdays off, using them mostly to tour the great monuments, museums, and buildings in and around Washington, without many of the weekend tourists. I remember going to the Museum of Natural History to see the Hope Diamond. That was a popular thing to view at the time and the lines were short during the week.

Our company participated in what they called Prelude to Taps in the spring. Our company performed a series of PT type exercises with our rifles (M-1's in those days), with chrome bayonets affixed. I remember one night First Lady Jacqueline Kennedy was there with Bobby Kennedy. The President couldn't make it for some reason. I had seen her often in ceremonies and three times close up at cordons that we did when they flew into Friendship Airport with Air Force One.

Our platoon leader was Lt. Samuel R. Bird. If you never met him, you missed out on something special in life. He was special in that he could reach out to the enlisted men as well as his fellow officers. I remember one day in early June '62, in the late afternoon, he came around outside the barracks where some of the men would sit on the grass on the parade ground and unwind after a long day. He said to one of our men, "I hear you like to wrestle Smitty. How about if I give you a try." And they did. They rolled around the grass for a half hour until neither one of them had any strength left. Then they hugged and promised to do it again. Lt. Bird came back a few more times, but I'd say it was always wound up a draw. The lieutenant went on to play in the most important public stage of his life, when he was selected to be the Officer in Charge of President Kennedy's casket team. Who can ever forget the picture of him leaning forward to help the men carry the casket up the steps of the Capitol? With the whole world watching, he sensed the men in his Casket Team needed help in keeping the casket level while they climbed all those steps with the tremendous weight they were bearing. I believe his greatest personnel triumph came after he was shot in three places on his body in Vietnam on 27 Jan 1967. He was almost completely paralyzed for the rest of his life, which ended 18 Oct 1984. His name was added to the Vietnam Veterans Wall. His wife wrote a book of his life named, "*So Proudly He Served - The Sam Bird Story.*" If you can find a copy in your local library, please read it and you will never complain about your little aches and pains after you read what that man went thru for his country.

A Company played a small but proud part in President Kennedy's funeral. Forming a cordon at various places along the caisson's procession thru the Capitol. I remember how cold it was, and the wind was blowing right thru our winter ceremonial outfits. But all of us collectively endured this brief period of discomfort because of the magnitude of this part of history that we were living.

I remember in the hot summer of 1963 when Martin L. King came to DC to have a solidarity rally at the steps of the Lincoln Memorial. There was much tension on the part of the civilian and military authorities for fear of an uncontrollable riot breaking out in our Nation's Capitol. We were put on high alert for what has become known as Dr. King's "I Have a Dream!' speech. As most military men know, while in the states, a military man very rarely is entrusted with live ammo even while on guard duty. For this occasion they flew in helicopters that landed on the parade grounds at Fort McNair. We then were issued live ammo for our weapons. I had a machine gun assigned to my squad. I had been made Sgt E-5 by then. Our objective to protect was the southwest corner of the White House. If the people started to mass outside on Pennsylvania Ave., they were going to fly us in. Our orders were not to let anyone come over the fence into the

Courtesy of T. Golden.

White House grounds. As we sat in and around the helicopters that hot day in full combat gear, with loaded rifles, we talked about how difficult it would be if we indeed had to shoot our own people. I mean shooting the enemy in combat is one thing, but in peace time, in the U.S., your own people? I'm glad it ended peacefully. It was such a great relief to clear our weapons that day and hand in our rifles.

We went to Fort A.P. Hill twice a year for refresher training because we were still in the Infantry. That usually led to a lot of stories and griping from the men. I remember one night we were supposed to be on a map reading exercise. I had my most experienced man study the map in the day time, until he was sure he could find our objective. Well, right after chow and when darkness fell, it started raining. It was cold and miserable. A truck took us out on the base somewhere and we were supposed to find this place on the map. They assigned an officer with each squad to grade you and make sure we didn't get lost or in trouble. After wandering around in the woods for the better part of an hour, my B team leader was hopelessly confused and so was I. The lieutenant became angry and said he would take over to get us back, but we would get a failing grade and have to do it over the next night. He took control of the squad and started off thru the woods with the rain never letting up and we were soaked to the skin and so cold we couldn't feel our hands anymore. After what seemed like another hour, the lieutenant called me forward and said he was lost. He had completed Ranger School and never had a problem with map reading but that night his experience failed him. He made a deal with me that he would call on the net radio to have a truck pick us up to take us in to the tents. In exchange I was sworn to secrecy to never tell anyone he was also lost. I said that would be ok if he gave us a passing grade so that we didn't have to do this again tomorrow night. He agreed. The next day he came down with such a cold, he had to go back to McNair for bed rest.

I was always very proud of my casket team members. We tried very hard and gave the deceased the respect they deserved. One of our members was so very terrified of the dead, he was so gripped with the fear that eventually, something would happen. Well, one day at Arlington Cemetery, one of our seven funerals that day, had just four people attending so they were pretty far behind us as we made our way to the grave opening. The casket was one of those Government Issue ones that are made of thick cardboard I believe. I always positioned this member of our team close to me so that I could help or calm him. All of a sudden I seen his eyes grow as big a saucers and he started shaking. I asked what was wrong. He said something from the casket just hit his leg. We were far enough ahead so that I could look under the end of the casket where I seen the casket floor was starting to come apart and the deceased hand had come out and hit this man. I told him to take it easy it was just the bedding that touched him. We were right by the gravesite by then so while the men paused there, I went around my man and tucked the deceased hand back inside and we laid the casket down. I didn't tell him what it was until we were back to the barracks. He almost fainted sitting down.

One more story on this same good soldier. After I begged and pleaded with him to stay on the team, telling him how we needed him, and he was a part of a team and couldn't quit and that nothing would ever happen again, he stayed. About a month later we had a funeral at Culpeper, VA. If you have ever gone to the cemetery there, you know it's up a fairly steep hill from the main road. That morning it snowed about three inches. Just enough to make the road to slippery for the hearse to navigate the hill. We had to carry the casket up the hill and with our leather soles and the metal cleats that are on the bottom, we weren't much better off than the hearse. The biggest problem I could see as we were going up the hill trying to keep the casket level and keep from falling, which would have landed the casket on top of us, was that I couldn't see exactly where the edge of the grave was because it was prepared the night before when there was no snow. The cemetery custodian always is there and he was walking along side of me. I said to go slightly ahead of us to define the grave. He said to follow him and he wouldn't let us get to close to the edge. Well wouldn't you know it, my old friend from the broken casket, who was right next to me, slipped into the grave with one leg, how we kept the casket from falling I'll never know? He screamed out, "Help me Sarge, Oh help me!" He was already crying and for a moment I thought someone was pulling him in there, because as much as he tried to extract himself from the prospect of going all the way in the grave, he kept slipping on the snow and was losing the battle. The custodian finally grabbed him. Luckily, the procession was having as much trouble as we were having and were far behind, trying to negotiate the ice and snow. Needless to say, my man quit after that.

Most of my memories are very fond of the time I spent with the Old Guard and I wouldn't trade them for anything. I know my fellow soldiers would say the same. I left them 9 Jan 1964. The Old Guard has formed an alumni Association, of which I am a member.

My Stories

by William Alva DeFreitas

I arrived at Arlington Cemetery from Fort Jackson in the afternoon. I was put downstairs in a holding with two other guys. I had no idea of the adventures waiting for me.

The next morning, I was assigned upstairs in the old barracks of Company E. From that

day on, I learned the high expectations of the Old Guard 3rd ID and took part in all of the ceremonies and drills.

I took part in many ceremonies and many burials because it was the Vietnam Era. One story will always stay in my mind and in my heart. I was at a burial when my eyes fell on the young widow, a thin, young blond woman holding her newborn baby. I recognized her immediately. She was the wife of the lieutenant I served under. He had left to go to Vietnam just two or three months previously. I closed my eyes and took a deep breath. I needed to retain my composure. I will never forget that day.

Other memories flood back to me - I was on duty for several parades and performed ceremonial duty for King Olaf of Norway, the Emperor of Ethiopia, the Argentinean and Brazilian presidents. I stood at attention in the hot sun as their anthems played. They were the longest National Anthems ever!

I was also on duty inspected by President Johnson and all of the big generals at the Pentagon. I was so proud. Here I was a foreign-born American, known in Company E as "Wet Back" standing proud for my country.

I remember one freezing night on duty in front of the Jefferson Memorial. They forgot about me! My replacement came in the early morning hours the next day. Finally, I was given the morning off and asked to report on duty in the afternoon.

Around "Easy" Company

by Cpl. William Feldman

• After a tough week at A.P. Hill Military Reservation in Virginia, the men of E Company returned to Fort McNair to find that their bunks had taken on that "new look." All the bunks had been painted a shiny white.

• The field problems were made a little more realistic with the help of Lt. "Puson Pete" Tyler and his gas grenades.

• While making an attack on a mock village, Sgt. Hershey's squad was captured. But that didn't stop the indomitable sergeant. He infiltrated right on through to capture the village single-handedly. Great work, sergeant!

• The members of Easy Company, due to the wishes of our 1st Sergeant, M/Sgt. Komoroski, have "volunteered" to get GI haircuts. Their new "bowling ball pates" give the men a proud feeling for it shows that each man can carry out an order – whatever it may be! *Submitted by Robert L. Hershey.*

Special Duty

by Robert A. Finton

Being 3rd squad leader I was low-man on the totem pole and drew Saturday duty for 2nd Platoon, D Company. My squad and I were tasked to assist some kids as they practiced marching with S&Ts (state and territorial flags) at Galludet College in Washington, DC. This was a major bummer as the weekend promised to be a beautiful taste of the coming summer of 1983.

Upon arriving at the school's athletic field with a deuce (2-1/2 ton truck) loaded with S&Ts I winced as I discovered that the "kids" were actually participants in the Special Olympics. My squad off-loaded the heavy flags and poles as I spoke with the cordial and engaging woman in charge. At the time I had never heard of Eunice Kennedy Shriver and had no clue as to her connections. I shared my concerns regarding the weight of the flags and the fact that any hint of breeze could whip a pole out of a Special Olympian's hands. The enthusiasm of Mrs. Shriver and the kids was undaunted and I guaranteed that my men and I would do our best to help.

Opposite: *Firing Party from 3d Plt., Comapny C, 1977. From left: SP4 Fullam, PU3 McConnell, SP4 Lindeman, Sgt. Adams, PFC Sabo, Sgt. Capps, SP4 Heinley, PFC Cooney. Courtesy of Dean R. Lindeman.*

These special athletes enjoyed the whole process and paid very serious attention to the practice. They really got a kick out of wearing the white canvas flag harnesses and jockeyed for positions to carry their own state's flags. I was tickled seeing my guys identifying state flags and shifting kids in the formation. The young marchers beamed with wide and bright smiles as they hoisted up banners from where they lived. But this was nothing as compared to lunch. My pride in my fellow Old Guardsmen soared as the guys declined to take their break separate from the kids. During their downtime, in the heat of a Saturday afternoon, soldiers of the 3rd U.S. Infantry played Frisbee and duck-duck-goose with Olympians.

As the practice drew to a close Mrs. Shriver expressed her thanks and hugs-galore were traded with my squad and me. We headed back to Fort Myer thinking that our goodwill mission was accomplished and the next weekend could be spent away from fatigues, blues and The Old Guard. That was until I got called down to the CO's office the following week. Captain "Mac" informed me that my squad and I were to be at an up-coming White House function. A bit concerned, I told the captain that my White House security clearance hadn't been approved yet. He smiled, handed me a piece of paper and said, "You don't need clearance. You're invited."

On 12 June 1983 my squad and I marched alongside a host of kids celebrating their Special Olympic achievements before President Ronald Reagan, This was also a day that I saw a side of President Reagan that made him even more of a hero in my eyes. As he stood at the lectern with the Presidential Seal displayed to the front, one of the Special Olympians in the gathered crowd ran toward him with outstretched arms. The fellow was a giddy and awkward child nearly thirty years old. As a Secret Service agent moved to intercept the young man, President Reagan waved the agent aside. The Presidential remarks in the Rose Garden were briefly interrupted as the leader-of-the-free-world knelt to hug and chat with a disheveled, but very special, athlete.

After the parade of flags, ceremonial presentations to the athletes, and remarks given by the President of the United States of America, we were encouraged to partake of the refreshments and join-in with the small talk. This White House festivity also involved the promotion of a newly released Superman film. Besides hob-nobbing with President Reagan and his wife, Nancy, my men and I met with Christopher Reeves, Richard Pryor, Susan St. James and other Hollywood types. Horrors known and unknown flashed before my eyes as my guys pleaded to be allowed to have a drink from the White House's open bar. "Okay. But one, I repeat, one drink each," I replied with foreboding. Having been a sergeant for only six months I could not bring myself to relax and drink alcohol at the White House. Since we had to stay to retrieve the S&Ts after the function ended and the guests had departed, my guys were in position to be offered table centerpieces and remaining opened bottles of champagne (and some bottles not so opened). I was about to be the wet blanket until a Secret Service agent chided me with, "Go on, Sarge. You guys were guests and worked to boot." Needless to say, I threatened severe pain and general bodily harm if any one found out about these souvenirs. But I had to chuckle. My men impressed me and made me proud to be an Old Guard squad leader with the attention given to and the time spent with our Special friends.

Eyes Left

Following is a letter Robert Haney received from then 2nd Lieutenant Frank Norris.

The bus was late due to the Memorial Day holiday. We spent the previous day and evening putting out those little flags on each grave in the cemetery. The civilian driver was supposed to have been briefed as to the location of the staging point for the parade. He had not.

We spent a very uncomfortable morning attempting to find our destination in our dress blues stuck in the holiday traffic. We finally found a parking place somewhere near our departure point, walked single file (with weapons—again, in dress blues) to our staging point which was cleverly positioned between two high school bands.

As the parade began, both bands played different tunes—very helpful to cadence. Fortunately all in the "Third Herd" were so well trained that this made little difference.

After about 2 miles of harmonized humming of "*I Love A Parade*," we approached the reviewing stand. The idiot civilians had put it on the left of the street, which contained the honored guests and Colors. Under arms, a unit cannot pass Colors without rendering honors. This is lovingly known as "eyes right" while passing in review.

The following is a narrative of that morning's fun and games taken from a memory:

My guidon, the sergeant directly behind me, SGT Robert Haney, whispered, "Sir, I think we have a problem.

After the joint "oh shits" I quickly utilized my highly honed skills fresh from OCS, I whispered back to Haney, "Do you think you people can pull off an "Eyes Left" if I call it on the left foot." Since this particular footwork is not found in FM-22-5, Drill and Ceremonies, there were no provisions for such a command ("Eyes Right" is the only approved ritual) and as such, never practiced, performed or best yet, approved.

Haney whispered back to the platoon sergeant (to his rear, front row) and the reply, "If the Lieutenant can get it right, we can."

As we approached the reviewing stand, I gave the command "left shoulder arms" (we switched shoulders during the dance as it involved twirling and a lot of hoo haa—very impressive if performed by a well trained

"Flags In" at Arlington National Cemetery.

unit) and then "eyes left." The platoon performed the command (in actuality in reverse of the normal ritual) without a miss and we marched on our merry way.

I don't think this had ever been performed (certainly not by the Old Guard) it was a first. Thank God the Battalion Commander never got wind of this - if he did he probably was amused, as the civilians there at the parade would never have known the difference.

Bob, if you remember Arlington Cemetery, you will remember that the access roads resemble a spider's web—a very large spider's web. As a new butter bar (2nd LT) I had no idea how to get to any of the gravesites as we lead a casket, the US Army Band and friends and relatives through the "Bone Yard" on funeral jobs (drops). You were always there behind me whispering instructions to the "new-be," thus keeping me out of trouble, especially when it was raining or snowing. We averaged five jobs a day during that time and it was nearly impossible to recon the route prior to the job—remember that we got the drop sheet on the morning of the jobs. It would have been very embarrassing to parade around the Bone Yard looking for a gravesite. Can you imagine what the crowd behind us would be thinking as they passed by a very familiar landmark for the second or third time? How about the multitudes of visitors that were everywhere?

Bob, you are an excellent example of the kind of people who make up the Old Guard and I thank you for your patience and very thoughtful help throughout that year. I hope to see you again.

Frank Norris
MAJ MI (USAR Ret.)

Recollections Of A Former Fifer

by Thomas Hanson

Inauguration Day January 20, 1981

I woke up excited about the day. Ronald Reagan was about to succeed Jimmy Carter as President of the United States. The Iran hostages were still in captivity. "Woody," our Drum Major, tied a yellow ribbon - the national symbol of unity for the hostages - to the espontoon he would be carrying in the parade.

During the bus ride to the Capitol, where the escort would begin, the news came over the bus radio that the hostages had been freed We all cheered and applauded, "Woody" untied the yellow ribbon, opened a window, and set it free in the wind.

The Longest Day - October 19, 1981

It was 200 years to the day from General Washington's victory over Lord Cornwallis at Yorktown. 10,000 troops were in formation, a million spectators were on hand, and tens of millions more watched on TV. President Reagan spoke, President Mitterand of France spoke and a congessman from Virginia also spoke. All speeches were translated for the English and French-speaking audiences. From the time we came to attention, marched into position, stood at parade rest, passed in review and marched back to the buses we had been in formation for over six hours. Troops were dropping like flies all around us.

After we had troop-stepped past the reviewing stand without a single hole in our ranks, our Corps Sergeant "Jake" Octavec said, "I'll tell you what, I was proud "

One Eternal Moment- February 5 1982

The day I left the Army, "Woody" Woodwell gave me a ride to the airport. I had about a half an hour before boarding time so I walked outside and gazed at the Washington Monument looming on the horizon.

As the plane climbed and banked I found myself looking directly down at the tiny specks of white over the green grass of Arlington National Cemetery. I found Fort Myer, a small semicircle hugging the west wall of the cemetery and spotted a little brick building by the back gate that was Company H. I thought, "How could a place so small have been such a big part of my life for two years? "

In one eternal moment I saw my entire time in service flash before my eyes. I was both happy and sad. Sad that it had come to an end and happy that I had done my best. I sat back in my seat and thought about reuniting with my family and friends in Minnesota. I had so much to tell - and still so much to learn. "But, " I thought, "someday I'll return.

This Sacred Land

The Old Guard's western wall
Inside, the flag unfurled
Protects the tradition of honor
From the noisy outside world

Lush green rolling hills
And row upon row of white
Slope gently toward the river
It's a silent, solemn sight

The silence ripples outward
From this Potomac point
it echoes through the nation
Our citizenry to anoint

The anointing is of peace
And an unwavering stand
To guard this place of honor
To protect this sacred land

When Eagles Cry

Eagle's screams
America's dreams

Sail to the wind
Nation's colors rippling

An eagle cries
When a soldier dies

Glide over a breeze
Watch over our seas

With keen eyes
Always on the prize

The Inaguration of President Ronald Reagan, January 1981.

A Promise Kept

Waiting as the drizzle pours
Just outside the chapel doors
The Army's best
Stands at parade rest

Restless horses
And mourning people
The church organ's tones
Drift over the steeple

The caisson is loaded
The casket secured tight
He's one of our own
The burden is light

Horses' hooves on cobblestone clatter
It's pouring rain, but it doesn't matter

To the garden they march
At a stately step
A life well-lived
And a promise kept

A Moment To Remember

by James T. Hawthorne

I was stationed at Fort McNair, Washington, DC when Ronald Reagan was the President of the United States and George H. Bush was the Vice-President. Occasionally, Vice President Bush would come to Fort McNair to do his jogging. Usually, we would be told in advance of such visits so as to prepare us for the additional security that was to be expected for such a visit.

A group of us were out in front of our barracks preparing to leave for parade practice at Fort Myer when the Vice President of the United States approached at a double time with his Secret Service detachment trailing nearby. My friends and I got a kick out of the Vice President's workout attire. On the front of his blue T-shirt were his initials "GB" in large print. Above them in smaller print was "The Vice Squad." As he jogged by, we noticed that on the back of the shirt was "Leader 2." It all happened in about 30 seconds or so, but looking back on it, I have come to appreciate the moment very much.

It was the possibility of experiencing just such an event that attracted me to the Old Guard in the first place. My only regret is that I didn't say at the time, "Mr. Bush, will you please join us for a photo?"

The Longest Night Of My Life

by J.B. Conmy Jr.

"Sir, the Lippanzaner Stallion has just jumped in the Tidal Basin and we need your sedan to search for him," shouted Lieutenant Paul Redner as he came running toward me. It was very apparent from his highly excited state that this shocking revelation was indeed true. In a flash, Lieutenant Redner's alarming news transformed what had been a most relaxing and enjoyable evening for me into an ordeal that was to be the longest night of my life.

My command, the Third Infantry, the Army's elite unit responsible for national military ceremonial activities in the Washington DC area had just completed its weekly summer evening military pageant for tourists at the Jefferson Memorial. Most of the audience had dispersed and personnel of the Unit were busily dismantling and loading equipment for return to our base at nearby Fort Myer, Virginia. The magnificent Lippanzaner Stallion, used during the pageant, was a gift to the United States from the Government of Austria, in appreciation for General George Patton's rescue of the Lippanzaner herd in the closing days of World War II and its return to Vienna. The Third Infantry, in its ceremonial capacity and the only military unit still authorized horses, was selected to receive the Lippanzaner named Conversano Beja. As Beja was being loaded aboard his darkened van, he spooked, pulled the halter from his handler and bolted into the darkness, little knowing that he was headed for a swim.

The thought of the international implications if this valuable piece of horse flesh was lost and the probable consequences to me as a career Amy officer raced through my mind inspiring me to do something right now! I quickly assembled the few personnel present, which included my wife and eleven-year-old daughter both of whom responded to my suggestion that they go home as it looked like a long night by saying, "What, and miss all the excitement." I hoped I would have better luck in finding the horse.

Col. Alois Podhajsky, Director, Spanish Riding School, Austria, Vienna, presents "Conversano Beja" to the U.S. Army at Madison Square Garden, New York City, NY, May 6, 1964. Courtesy of Michael D. Juers.

Our group broke down into three search parties, two by land, circling the Tidal Basin in opposite directions, and one by "sea," aboard a couple of the Swan paddle boats normally used by tourists for pleasure riding on the water. The amphibious element was quite a sight to behold. As these military personnel mounted on paddle boats, garbed in full dress blue uniforms peered with hand to brow, into the inky blackness of the Tidal Basin. The land patrols soon became most unpopular with lovers occupying park benches on the fringes of the Tidal Basin who responded with unprintable replies as they were asked "Have you seen a white horse pass by here?"

As the search wore on I became more disheartened until out of the darkness once again loomed Lieutenant Redner with the report "Sir the Swan Boat Patrol has located Beja and herded him against the sea wall near the Swan Boat Dock." I heaved a sigh of relief as I headed toward the rendezvous point. When I looked over the three foot high iron railing above the sea wall I was greeted by the sight of Beja up to his neck in water with Chief Warrant Officer John McKinney and Lieutenant John White, dressed in colonial costume and full dress blues respectively, clinging to Beja attempting to quiet him. Beja was in no mood to be comforted as he lashed out with his hooves. To further complicate matters there was no ramp exit from the Tidal Basin and the tide was rapidly rising. It was apparent Beja would have to be lifted out of the water and no evident solution was at hand.

It was now well past midnight and it appeared we would avoid a crowd. I had no sooner had this thought than over 500 people seemed to come out of the woodwork to give us the additional burden of crowd control. I was congratulating myself that no press had appeared when up stepped a young woman reporter, complete with photographer, from a major DC newspaper. Her first question was, "Colonel, what is happening here?" It was immediately apparent to me that my initial reply of "I don't know" was not going to satisfy her; for with that response she surveyed the front of my uniform and said, "Lets see, you are a Colonel, you are in the Third Infantry and your name is Conmy." I promptly replied, "I surrender" and told her of our plight for which we had not reached a solution at the moment. All I could visualize at this juncture was my forthcoming trip to Vienna to explain to the Austrians what had happened to their horse.

With the arrival of the crowd, police began to appear. We soon had representation from the U.S. Park Police, DC Metropolitan Police, Arlington, Virginia Police and Armed Forces Police, each with a different idea of how to get the job done. They wanted to be helpful but the blare of their radios merely caused Beja to thrash around resulting in more barked shins for my unfortunate cohort down in the water. At this point I decided I needed some professional help and put in a call for the Military District of Washington (MDW) Veterinarian.

While waiting for the Veterinarian, I requested the DC Metropolitan police to get us a tow truck as a device to lift Beja out of the water knowing that at 2:00 a.m. only the police would know where to find a cooperative service station. The tow truck arrived at the same time as the MDW Vet. Maybe now we could make some progress for the rising tide made each minute count. It was soon apparent that the tow bar of the truck would not reach far enough over the water to lift the horse so we must find another solution and fast. However, the Vet did arrive with a very useful piece of gear, a canvas and leather sling that must be strapped under the horse's belly to provide a means of attachment for the lift to be accomplished. So the vet and his two assistants joined the melee in the water for the sling strapping operation. As the tide continued to rise and I observed the struggle in the water below, I figured my next call might have to be for Navy divers.

When it was obvious that our service station tow truck could not handle the lift task, a Park policeman suggested the DC Transit authority's tow truck which was used to tow heavy buses. by the time this truck arrived it was after 3:00 a.m. but again we were doomed to failure as the tow bar was not of sufficient height or length to reach out over the water to effect the lift.

In the meantime the veterinary team had the sling secured around the horse's belly and my daughter, Mary Alice, was hanging over the wall feeding Beja sugar cubes she had located at a nearby all-night restaurant, which at least quieted Beja and my comrades in the water suffered fewer kicks in the shins. All we needed now was some way to lift that 2,000 pounds of horse flesh.

Our last resort was the 5-ton wrecker at the Fort Myer motor pool. Having used it recently during our field training, I recalled it had a long telescoping lift arm and was high enough to reach out over the water to permit attachment of the sling. With the aid of the Armed Forces police we sent word to Fort Myer and the wrecker was dispatched to the scene of our dilemma. When the wrecker arrived it worked just as I had surmised it would.

At 4:15 a.m. the lift hook was attached to the horse's belly sling and Beja was lifted from the water amidst the cheers of our large audience.

I breathed a silent prayer but my relief was short-lived. As the wrecker lowered Beja to the ground all four legs buckled and Beja rolled over on his side giving us all a very glassy-eyed look. Once again I looked in the direction of National Airport wondering what time the next plane left for Austria as I was sure all four of Beja's legs must be broken and we would have to shoot him. However, the vet quickly checked Beja and informed me that, "Beja is in shock, I am sure I can bring him around with a couple of injections" which he promptly did. As Beja staggered to his feet I mentally cancelled my plane reservations for Austria and only hoped that the press would be kind.

Fort Myer never looked so welcome as we rolled through the gate with horse in tow. Once in his stall Beja was thoroughly examined by the vet and found only to have suffered a small puncture wound in one leg, which was dressed and he was as good as new.

As I left the stables and walked toward my quarters, the sun was appearing on the horizon, thankfully bringing a happy ending to the longest night of my life.

A Positive Performance

by Tom Marcotte

In the early 1980's two books about B Company 2/3 Old Guard were published. I took exception to some of what authors had to say. In passion and anger I wrote the following, and then put it away. From time to time I'd get it out and re-read it. Over time the anger subsided....the passion did not.

The 199th Light Infantry Brigade was a composite of three line battalions and various support units. It had, until 1969 spent its tour in the Mekong Delta South and West of Saigon. The battalion in which I served was the 2nd of the 3rd Infantry. The 3rd Infantry is the Old Guard which serves as the unit responsible for guarding the Tomb of the Unknown Soldier at Arlington National Cemetery.

During Tet of 1968 the Old Guard had fought in downtown Saigon and done well. Following that period the 2/3 returned to the Delta and engaged in pacification. The VC elements in the area were exhausted from Tet 1968 and by the time I hit the field to join Bravo Company in late December 1968, war in the classic sense of engagement and battle did not exist. The real battles were with boredom, drugs, venereal disease and booby traps. Many of the guys who served from the spring of 1968 till the spring of 1969 never fired a shot at the enemy.

It wasn't for lack of trying but in the vicinity of the Fishnet (2/3 HQ) there was no war. The VC were active but the tools of their effort were booby traps, drugs, prostitution and planting the seeds of dissent between the races and generations in our Army.

In May of 1969 the 2/3 was sent to Cu Chi for a relief of the Wolfhounds of the 25th Infantry. Here the 2/3 engaged in one major encounter with the enemy remarkable only in that the battalion commander, a man of limited infantry experience ordered the battalion to on-line assault against a dug in enemy. Surprisingly there were few casualties except the battalion commander who for our sakes was mercifully relieved.

We did foray into the HOBO woods and had our first taste of patrolling in an area where the enemy was prepared to do more than booby trap. We did not however engage in any major combat.

In July of 1969 the battalion gave up the Delta war and displaced to the Xuan Loc district north and east of Saigon along Highway One. For what we found here we were totally unprepared. The Michelin rubber plantations and surrounding jungles were full of NVA...battle tested and committed to our destruction.

Accustomed to operating in and among civilians in wide open and pacified rice paddies, the jungle proved a mystery. We hacked noisily through the steaming semi-darkness lost and naïve'. The enemy left signs of his presence but seemed no more willing than his southern VC counterparts to stand and fight...or so we believed.

The object of our long search was MR7: NVA Military Region 7 Headquarters thought to be in the area of Xuan Loc. Its HQ we were told consisted of an NVA Major General and its support complement. Our prize was the destruction of that headquarters.

We hunted the enemy with no concept of what would happen if and when we found it. We did not know what to expect. If ignorance is bliss than B Company should have been a very happy group...but no one was smiling.

In turn, each of the Battalion's Companies staggered into semi-destruction. Delta company found MR7 and was nearly wiped out. Particularly demoralizing was the crash of a Medivac chopper into the company command post when hit by a misguided eight-inch artillery round.

The NVA entered the radio net and talked to the dying remnants of Delta. We could listen to them on our radios but somehow didn't hear.

B Company at this time was commanded by a captain who was on his second tour in

Vietnam. This special forces officer was well liked and respected by the men who recognized that he most wanted us to go home having done our duty…alive. I think that he already knew what others were to realize only later…Vietnam was going in the loss column.

The action of September 5, 1969 in which B Company engaged MR7 was a battle, not a firefight. It lasted over six hours, leaving six dead, twenty-two wounded, and a lifetime of unanswered and unanswerable questions for those who were there.

The bad luck for B Company began with the departure of the balance of promotable E-4's and E-5's to the rear for a promotion board. Having completed a BDA…Bomb damage analysis after a B-52 mission and finding nothing; the LZ left by the destruction proved more than suitable for the removal of the key junior leaders. The loss of their experience and talent would be critical in the coming hours. The agitated departure of the Hoi Chan scouts who left at the same time should have also given a clue that something wasn't right. Finally we had never been told that a B-52 strike was never staged when less than a battalion of enemy was the target.

Under strength, with something less than 65 men, lacking in leadership in critical positions and ignorant of danger, B Company walked north off the moon scape left by the B-52 towards the reinforced ambush positions on MR7.

Third platoon was in the lead, my platoon, first was second, the company CP followed and second platoon was last in file. We had gone less than 1000 meters skirting a small swamp when we entered an ever-thickening jungle. As third platoon moved slightly up hill I was three back from the tail of third platoon's formation and 50 to 75 meters behind the company point.

The radio traffic between the third platoon leader and his acting lead squad leader indicated that there seemed to be signs of the enemy about and an odd smell in the air. The exchange continued as to whether additional point activity was required or should the company press on.

The next sound was the horrendous explosion of Chi-com claymores and rigged unexpended artillery rounds in the killing zone the squad occupied. The nearby trees exhaled bursts of sniper fire and B Company began its baptism of fire.

Third platoon was combat-ineffective momentarily and their platoon leader was asking for help. As the next platoon in line I asked and was directed by the company commander to move forward to

Members of the Old Guard prepare for repositioning in Vietnam.

assist. I cannot forget the eyes of those faces in the platoon asking me what are we to do.

I was 21 years, had been in the Army less than 18 months, had been an NCO less than a year, if in fact I was even entitled to call myself an NCO at all, and had never been in combat. Even to this day my reaction then and now still suprises me. I jumped to my feet, dumped my ruck and in astonished anger said, "lets go, those sons of bitches are killing our friends." Some very good prior training or too many John Wayne movies, I'm not sure which.

What I am sure of is that up until that moment I had seen little blood or mayhem and was still mostly detached from the reality of war. In an instant it all became very real and very personal. I am also certain that to a man all the men of 1st platoon ran up that hill with me. They were willing to try and did not lack for courage in that moment.

Later events would diminish each man in his turn and some would do less or more than others. All men are not created equal. As some are bigger or faster or smarter so also are they capable of greater or lessor measures of courage. To do your duty is to give what you have as long as you have it. Courage is not constant but ebbs and flow in and out of each man like the tide. And it is cumulative. The small brave act of one man can rekindle the faint hearts of many. The power to go on does not fully reside within but draws strength from the group.

I found both courage and cowardice in others and in myself. It is neither kind nor wise to be quick in judgement on those things not fully understood. The horror visited on B Company men that day took each man to his limit and each must reconcile with God and no one else. I believe that in the balance the men of B Company acquitted themselves honorably and nobly and do not deserve the unkind slander which others so callously infer.

The only sure way a combat veteran can know he has done all that he could is to be dead. Combat survivors will find themselves feeling guilty for the rest of their lives for just that - being alive when other are not. But for B Company the record of achievement that day is posted for all to see. Criticism and doubt casting cannot, should not, and will not change it. Sixty untested infantrymen fought a battalion-sized force to a stalemate, and all but a few lived to tell about it.

Three Silver Stars, a Distinguished Flying Cross, five or more Bronze Stars and most importantly 28 Purple Hearts testify to the integrity of the men of B Company.

Even as you read this, solders of the Old Guard still quietly parade at Arlington. They and all who came before them and all that will follow them can be sure that the men of B Company 2/3 Old Guard did not let the banner fall.

Vietnam for many has become a dark night in a deep hole on a black and angry sea of confusion and emotion. And my heart is with them in their trials. For me it seems surviving the battlefield is only the first step in surviving a war. Later coming to terms with all that you have seen and all that you have done can be infinitely more difficult. The trick is not to try to understand too much too fast and to accept that some things can never be understood at all.

The final words of anger, frustration and doubt will all be still in time. Among those who will speak of Vietnam as only history I would ask that they be gentle with the truths so meaningful to those who fought their way through them.

The "Out to Lunch Bunch" A Whimsical Tale of Barracks' Hi-Jinks

by Brett Reistad

Among my fond recollections of holidays in The Old Guard is an incident that occurred thirty years ago, during the winter of 1974. I was about seven months into my enlistment, and a new arrival to "H" Company's Salute

Gun Platoon (Presidential Salute Battery). At the time we were billeted at South Post Fort Myer, Virginia (now a part of ANC) in the "Yards & Docks" building, which served as our home while our vintage North Post barracks were being remodeled into a dormitory-style layout more conducive to the new All-Volunteer Army. I roomed with two other "new dicks" who like myself, were learning their way around and assimilating to the unique conditions of barracks life.

Barracks dwellers from "Guns" and Recon Platoons shared the first floor - a mixture of enlistees, draftees, Vietnam vets and a newly arrived compliment of troops transferred from the 82nd Airborne Division. We worked hard and played hard together - some harder than others. Some of the more lively and mischief-prone characters stayed under the watchful eye of Platoon Sergeant (PSG) Frederick Jones ("I'm the Billy-Joe 'round here...), and became known among their peers as the "Out to Lunch Bunch."

As it happened, one evening I was rudely awakened from a sound sleep sometime after midnight by several of my colleagues, who obviously had not given much thought to reveille. They were in the process of mustering all available troops to build a snowman in PSG Jones' parking space, to greet him in the morning. I reluctantly dressed to go out and contribute to this undertaking, only to find it snowing heavily. My co-conspirators were already collecting a mass of snow, and laughing in anticipation of PSG Jones' reaction.

As the snow mass began to take form, someone commented that it resembled George Washington seated on a commode - this was all it took to fine tune the creation and add some O.D. spray paint before a couple of hours sleep. All along, we assumed that PSG Jones would pause upon arrival, marvel at our creation, perhaps mutter a few choice words and commence to flatten George with his full size car.

Unbeknownst to us, the snow turned to freezing rain, turning our creation into a huge block of ice. PSG Jones arrived on cue, reacted much as we expected, and commenced to drive-over George. The snowman, however, would not budge. Try as he may, PSG Jones was doing nothing but becoming more determined, and obviously angry. As we peered from our respective vantage points laughing uncontrollably, in one final effort he got a running head start and again plowed into George without effect. Jones exited his car and upon viewing damage to his bumper, became highly irate. He immediately stormed into the building, launched into a loud tirade and initiated a witch hunt for the insubordinates responsible. After unsuccessfully attempting to identify the culprits, he issued an ultimatum - the snowman would be removed immediately from his parking space, or the entire platoon - NCOs and all - would suffer his wrath.

Needless to say, the prospects of mass punishment were inconsistent with our objectives and the Out to Lunch Bunch wanted no additional scrutiny of their after-hours activities, so in the interest of our self-preservation, the parking space was quickly cleared.

Most Memorable Duty

by Gary D. Rogers

The period of history from 1961 through 1963 was a very difficult time for the United States military. During this period, relations with the USSR and Cuba, were at times, extremely tense, and along with many other individuals in the military, I experienced interesting times such as the period when the Old Guard was on battle alert during the Cuban missile crisis and we lived and slept for more than a week in full battle dress.

Also during this period the Civil Rights era was in full swing and our unit trained for possible riot duty in DC, which fortunately never occurred. However, our everyday duties were much as they remain for the Old Guard today: ceremonial duties around the nation's capitol. My assignment was with Charlie Company as a casket bearer. My casket team participated in many burials such as the one for civil rights leader Medgar Evers at Arlington National Cemetery. We also participated in the arrivals and departures of heads of state such as Haile Selassie, emperor of Ethiopia. We marched at Gettysburg during a commemoration of the Gettysburg Address.

By far, however, the most memorable ceremonial duty that I performed during my service in the Old Guard was the funeral ceremony for President John F. Kennedy. Returning from what I presumed was my final burial duty before processing out, we were met by our company clerk, who informed us that President Kennedy had been shot. During the next several days, all activity was focused on the Kennedy funeral. When President Kennedy's casket was removed from the Rotunda to be taken to Arlington National Cemetery, I was posted on the steps of the Capitol, along with members of the other services. Then I was posted as a road guard on the route of the funeral procession into Arlington National Cemetery. My out-processing was then accomplished in less than seven days, no doubt a record.

The defense secretary of the country of Columbia was scheduled to lay a wreath at the Tomb of the Unknown Soldier. The Old Guard carried the Columbian flag. The flag was covered in canvas. Fortunately, the colonel in charge of the ceremony checked the flag and discovered it had been secured upside down. An International incident was avoided with the use of a screwdriver that allowed the color team to reattach the flag in its proper position.

Although I was privileged to visit Washington, DC on several occasions during business trips over the years and also to visit Fort Myer, it was not until 1997 that I was finally able to take my wife to DC, and show her Fort Myer, the barracks, and the Old Guard Museum – I have always been, and always will be, extremely proud of my time with the Old Guard and my service with the Old Guard to this country.

A Sense of Humor

by Chris Sharpe

During the winter of 1993 I was part of an eight man casket team conducting a full honors funeral in Arlington. I forget the exact section this took place in but it was on the side of a steep slope, I believe near the Coast Guard Memorial. It was for a retired female colonel who was much esteemed and there were several retired general officers in attendance. It had recently snowed so we were worried about traction while carrying the casket down the hill and onto the gravesite. We decided that we would take our time setting it down on the rollers and pause for a second prior to letting go and grabbing the flag to make sure everyone had a solid footing. It was a little treacherous getting down the hill to the gravesite but we made it without any problems.

Once we were at the site we carefully set the casket down, paused for an extra second as planned and all let go simultaneously, much relieved at getting through it without any problems. As soon as we let go, we realized that the groundskeepers had forgotten to insert screws into the "rollers" that we had set the casket onto. True to their name the casket began to slide downward on the rollers. The casket continued sliding toward the bottom of the hill until the head end of the casket cleared the first set of rollers and the entire casket fell, quite noisily, headfirst into the hole with the foot end sticking up through the flag we were now holding. For the next several seconds (which seemed like an eternity) we stood in stunned silence, except for the gasps from the family members, until the Chaplain stated finally "Well, I hope she had a sense of humor." With that we passed the flag to one member, got on our knees and began to pull the casket back out of the hole. This would have been a difficult enough task without being dressed in the winter overcoats and the now pointless attempts at keeping a ceremonial demeanor. Finally the limo driver grabbed the end that was sticking up and began to pull it toward him, which gave us the extra leverage we needed to pull the rest of the casket up and back onto the not stationary rollers.

We completed the rest of the ceremony without incident, most of us considering what it would be like to spend the rest of our careers in Korea once the Colonel heard about this. Happily, the Chaplain spoke with the family who not only understood but were quite impressed with the way the situation was handled. As a matter of fact, that Chaplain put the entire casket team in for AAM's which we received several weeks later. Ten years later it still brings a small smile to my face when I see that award and think back to the achievement that it represents.

Opposite: *General Douglas McArthur was relieved of command by President Harry S. Truman in April of 1951. He and his family left the Orient on April 15th for Washington, D.C., where he gave his famous address to Congress. This picture was taken the day before that speech, when he spoke to a crowd in front of the Washington Monument. The General is at the podium, his wife and 12-year-old son are at the far left. A portion of the Honor Guard (the group on the left) were from Company A, 3d Inf. Regt. (The Old Guard). Shown left to right are Frank L. Mathieu, Gustaf A. Sponholz, Armand D. Venere, and Leland B. Frampton.*

Left: *Captain William H. Henry Jr., Commanding Officer of E Company, 3d Infantry Regiment, receives the Buff Stick Trophy from the MDW Commanding General, Major General Stokes. The trophy was awarded in 1955 to the "Best Company" in the Old Guard - the company amassing the best overall record in all events during the preceding year.*

OPPOSITE: M/SGT Alpha Price, WAC, Minneapolis, MN, salutes the American flag in tribute to the 176th Anniversary of the establishment of the USA and Flag Day. SGT Gerald Comer, Ray City, GA, in present day ceremonial garb offers a marked contrast to the color bearer SGT/1C1 George Osborne, Pineville, KY, wearing the authentic dress uniform of the army of 1775. Photo taken at the Pentagon building, June 12, 1951.

BELOW: While serving in Korea in 1951, assigned as 1SG of K Company, 15th Infantry, CSM Manuel J. Silva was honored with 2 Silver Stars for action during a two week period at Chokko-Ri and Chiro-Dong. On August 4, 1952, his company was assigned to assault a strongly fortified enemy position. After a brief, bloody contact and order to withdraw, he skillfully regrouped the assault unit, placing himself at the head, and crossed open terrain to entice the enemy to reveal their position. He then led his men and directed their fire. Upon breaking contact with the enemy, he remained behind to insure safe withdrawal of all. On August 15, assigned to assault an occupied hill near Chiro-Dong, being bombarded by intense artillery, he disorganized the men and inflicted many casualties on friendlies. While evacuating casualties, they were being overrun. He regrouped his unit and later led a squad, firing rifles and throwing grenades. During this action, he was seriously wounded by an enemy anti-tank grenade, but refused medical treatment and continued to direct the attack and assist in evacuating the wounded. Because of his heroism in two wars and outstanding record, he was selected to carry the 15th Infantry colors in President Eisenhower's Inaugural Parade and was assigned to "The Old Guard" as 1SG of A Company. He was awarded the Purple Heart for injuries in Korea on the Parade Field at Fort Myer, Virginia. Courtesy of Manuel J. Silva.

"The Old Guard" Association

CHARTER

By authority of the Executive Council, "The Old Guard " Association, be it known that in compliance with the current By-Laws, did make application to form an Association in Fort Myer, Virginia.

The number of members constituting the minimum requirements for the Election of Officers being eight (8), the following have been elected:

CSM Ret. Thomas Twomey PRESIDENT	LTC Mark H. Armstrong VICE PRESIDENT	SGM John J. Oktavec ADJUTANT	MAJ Tracy G. Bryant TREASURER
CSM Ret. George Otis DIRECTOR	CSM Ret. Fred Darling DIRECTOR	J.R. Pete Holder DIRECTOR	Col Ret. John J. Madigan, III DIRECTOR

The initial chartering members are:

COL Gregory C. Gardner	MG Donald C. Hilbert	BG Ret. James F. Hennessee
CSM Ret. Thomas Twomey	CSM Ret. Albert C Huddy	COL Ret. Stanley Bonta
CSM Ret. George Otis	COL Ret. Mark Brennan	MSG Randy Hauck
J.R. "Pete" Holder	MSG James Hicks	SGT Darren Blue
MAJ Tracy G. Bryant	MAJ Matt Halder	LTC Ret. William A. Lancaster
LTC Mark Armstrong	CPT John Couig	PVT George R. Otis
CSM Michael L. Bergman	CW4 Jeanne Pace	MG Ret. Richard M. Lee
SGM John J. Oktavec	Marianne (Jill) Weigand	PFC John Babcock
Thomas Hanson	CSM Ret. Fred Darling	1SG Theo Matuskowitz
MG Ret. Leo M. Childs	Donald G. Cofsky	CSM Robert L. Phifer

and having consummated the requirements of the Executive Council, "The Old Guard" Association, is hereby authorized to organize and constitute an Association in Fort Myer, in the State of Virginia, as "The Old Guard" Association, the Twenty-eighth day of March 1998.

In witness thereof, we have hereto set our hands and the official Seal of the Association, this Twenty-eighth day of March 1998.

Colonel Gregory C. Gardner
3rd U.S. Infantry "The Old Guard"
Commanding

CSM Ret. Thomas Twomey
"The Old Guard" Association
President

The idea for an Association of present and former Old Guard Soldiers first came to light during a casual conversation between then-Regimental Commander COL Gregory C. Gardner, and CSM(R) Thomas Twomey as they both stood in Arlington National Cemetery awaiting the start of the Veteran's Day program in 1997. This conversation, and a series of follow-on meetings between interested parties composed of former and Active Duty Old Guardsmen, led to the formation of The Old Guard Association (TOGA), the vibrant and healthy organization we now enjoy.

The first appointed Board of Directors and General Membership meeting was convened at Fort Myer, Virginia on 25 February 1998, where The Old Guard Association By-Laws were presented and ratified. The document was presented to Garrison Command, requesting authority to operate on the installation under authority of AR 210-1.

The Association Charter, listing the thirty Charter Members of the Association, was drawn on 28 March 1998.

The Association's stated Aims and Purpose were laid out in the By-Laws, to wit:

a) To establish a permanent organization that will further develop the common bond existing between all who served with "The Old Guard".

b) To maintain records and publish the history and achievements of the 3d U.S. Infantry Regiment and the "Old Guard Association".

c) To be of assistance and service in matters pertaining to Veterans matters.

d) To perpetuate the memory of deceased Veterans of the 3d U.S. Infantry Regiment (The Old Guard).

e) To maintain close liaison with the 3d U.S. Infantry Regiment (The Old Guard), Fort Myer, Virginia, and to recognize outstanding achievement by appropriate presentations.

Pursuant to paragraph "d" of the By-Laws, an early operational goal of TOGA was to design and dedicate a remembrance to all Old Guardsmen, in commemoration of their service to the Regiment. Funds were appropriated from the early member's dues payments, and the Board went about the business of obtaining site approval from Arlington National Cemetery administration officials. Concurrent with that stage, design options were explored and refined, and a bronze plaque was cast by a Virginia foundry.

The Directors recognized that staging a reunion of Old Guardsmen would serve towards meeting the Association's stated Aims and Purposes. In addition, a reunion would serve as a catalyst for building membership. A committee was formed to explore the options available, and a timeframe of 2 to 5 October 1998 was selected to hold the first reunion, so as to coincide with Regimental Organization Days events being staged by the Regiment. The Days Inn-Arlington was selected as the headquarters for the reunion.

The first reunion was attended by fewer than sixty members and guests, but encompassed many of the events and features of later and larger such gatherings. One highlight of the reunion was a mustering of all former Old Guardsmen present at the reunion on Summerall Field. After forming into a ranks, the detachment Passed In Review before the Regiment standing at Attention with bayonets fixed, on the field where so many had once stood.

A second highlight of the first reunion was the unveiling and dedication of the Old Guard Memorial Plaque at the flagpole immediately south of the Amphitheater in Arlington National Cemetery. The ceremony, held less than one year after the idea of an Association came to light, marked the first successful completion of a tenet of TOGA's Aims and Purposes as stated in the By-Laws. As part of the reunion, a General Membership Meeting was conducted and election of Directors held. The first of our annual banquets was held at Spates Hall, Fort Myer, Virginia, and the first TOGA Invitational Golf Tournament was played.

During ensuing years the Association has staged reunions that have grown in scope and attendance. Popular events were held over from year to year. New events, such as Potomac River luncheon cruises, visits to regional battlefields where the Regiment served with distinction, tours of The White House and the Smithsonian Air and Space Museum Annex, as well as other area attractions were scheduled and well-attended. A reunion highlight of recent years has been our attendance at Spirit of America, the Military District of Washington and Army pageant to our Nation, as seen through the eyes of it's Soldiers.

The tragedy of 11 September cast a pall over the already-scheduled reunion of 2001. The reunion committee established telephone contact with all pre-registered attendees immediately after the terrorist attack on our Nation. A great majority of those attendees opted to continue with plans to hold the reunion, in the spirit of, "if we cave in, the terrorists win" attitude. Some intended guests were pre-empted from attending by the national grounding of all aircraft during the period immediately after the disaster. However, some who held seats on aircraft thus grounded were successful in finding alternate means of travel, even though that would mean a hardship for them. The pervasive attitude during that reunion was one of, "Old Guardsmen do what it takes".

Over the years, thanks to the cooperative efforts of the Association and Regimental leadership, TOGA has instituted an Outstanding Soldier recognition program, one of the Aims and Purposes stated in the By-Laws.

Additionally, the Association has contributed funding to the Regiment's Morale, Welfare and Recreation program. This funding has aided our Soldiers in staging "Org Day" events, as well as providing financial assistance to junior-grade Soldiers with families during the annual Holiday Season.

Special appreciation programs for Soldiers who have been, or are about to deploy to Global War on Terrorism Theatres of Operations are also part of TOGA's operational mission as stated in the By-Laws.

TOGA's leadership continues to research and consider new and innovative ways to meet the first of the Association's Aims and Purposes as stated in the By-Laws:

"To establish a permanent organization that will further develop the common bond existing between all who served with The Old Guard."

Taps

In Memory of Our Departed Association Members

John J. Kenny	1999
SGM (R) Lacy McDougald, Jr.	2000
Leonard Pilhofer	2000
George J. Hazen	2000
Col. Mark F. Brennan	2001
Frank W. Jenkins	2001
Gene E. Ackley	2002
Henry D. Trivelli	2002
Warren R. Magnuson	2002
Col. (R) Ian Sutherland	2003
Fred Grindol	2004
Edwin Barwick	2004
Edison L. Dowdy, Sr.	2004
CSM (R) Robert C. Chilton	2004
Col. (R) Fortunato R. Priore	2004

The Old Guard Association President's Hall of Fame

Peter J. McDermott

2004 2005

Donald G. Cofsky

2002 2003

Brett P. Reistad

2001

No Photo Available

Albert C. Huddy

2000

Thomas Twomey

1997 1998 1999

The Old Guard Association
Executive Council - 2004

Peter J. McDermott
President

Thomas Twomey
Vice President

John J. Oktavec
Secretary

No Photo Available

LTC Tracy G. Bryant
Treasurer

The Old Guard Association
Board of Directors - 2004

Ray "K" Kirschenheuter

J.R. "Pete" Holder

Robert M. Thompson

Charles A. Willis

No Photo Available

Gregory C. Gardner

No Photo Available

George E. Otis

3d U.S. Infantry Regiment
The Old Guard Command Group - 2004

COL Charles L. Taylor
73rd Regimental Commander

LTC Christopher R. Parsons
Deputy Regimental Commander

CSM Harry L. Wimbrough
Regimental Command Sergeant Major

Honorary Colonel of the Regiment

MG Donald Hilbert

Honorary Sergeant Major of the Regiment

No Photo Available

CSM Albert C. Huddy

The Old Guard Veteran Biographies

Company C (Charlie Guard), 3d Infantry, commanded by Captain Mark Lewis, 1977.

SILVIO R. ANNUNZIATA, born in Worcester, MA on May 10, 1939. Enlisted in the US Army Sept. 20, 1961. Assigned to the 101st Abn. then to 1st Missile Bn., Military Police, Chun Chon, Korea. Served in A Co., The Old Guard, February 1963 to September 1964.

Served in the USMC and USMCR for 4-1/2 years as a radio operator/ infantry. Duty assignments included, MCRD, San Diego, CA; Little Creek Amphibious Training School; and Camp Lejeune, NC.

Married Lynne M. Rhodes and they have three sons, two daughters and three grandchildren.

Employed as a district sales manager for Cooper Industries.

AL-AMIN S. AS'SALAAM, served as Johnnie S. Peay, born Brooklyn, NY, June 6, 1945. Enlisted in the Army, Aug. 29, 1962; received his basic training at Fort Gordon, Augusta, GA. The only assignment: The Old Guard, Delta Co. 1st Bn. (Reinf.) 3rd Inf., Fort Myer, VA.

While serving in The Old Guard, he worked on the Prelude to Taps Project, for which he received a Letter of Appreciation (Commendation); honor of serving with his unit at West Point Academy; President John F Kennedy's Funeral Parade; later stood vigil on President JFK's Gravesite (before the gravesite was bricked over); while at JFK's gravesite some of the visitors were former 1st Lady Jackie Kennedy; US Atty. Gen. Robert F. Kennedy; US Senator Edward Kennedy; and TV Host Ed Sullivan. The honor of serving in President Lyndon Baines Johnson's Inauguration Parade and at the Inaugural Ball; honor of serving in Gen. Douglas MacArthur's Funeral Parade. While serving The Old Guard, received his GED from the state of Maryland; and, began classes at George Washington University.

After leaving the Army, met Robert F. Kennedy, while working as a junior executive at the front desk of the Manger Hotel. Ordained minister in 1966, and became bishop in 1967. Lost his baby brother, Douglass Franklin Peay, in Vietnam, July 11, 1967. Named to the Dean's List for Academic Honors for the Winter 1983-84 and Spring 1983-84 trimesters at Sojourner-Douglass College. Name and featured in the 1984-85 edition of The National Dean's List. Retired Supervisor, EAS-16, from the US Postal Service, May 4, 1998. He makes his home in Baltimore, MD.

ROBERT R. ATCHISON, born Bronx, NY in 1933. He enlisted in the Army on March 20, 1951, Manhattan, NY. Basic training Fort Dix, NJ, 2nd Inf. Indian Head Div., 9th Inf. Regt. Korea, December 1951. Received Combat Infantry Badge, Silver Star, two Bronze Stars w/Korean Service Medal, Wharang Distinguished Service Medal, United Nations Medal, National Defense Medal, Good Conduct Medal.

Other assignments from January 1953 to April 1954, The Old Guard, 3rd Inf. Reg., G Co., Drill Team Platoon, Fort Leslie J. McNair, Washington DC. Rank was staff sergeant.

He was married to Constance Layne from Woodside, NY at the chapel in Fort McNair. They have two daughters, Deborah and Lisa; one son, Robert C.; and 11 grandchildren.

Being discharged on April 12, 1954, he worked as a trackman for the Long Island Railroad; fireman, Penn. Railroad; accountant, business consultant, 22 years with the NYPD and retired as a detective. Later he worked as a licensed private investigator.

He became re-acquainted with S/ Sgt. John Murphy after 49 years, thanks to The Old Guard Association.

JAMES R. AUX, was in the active Army from 1970 until 1994. He served in Charlie Guard, The Old Guard, from 1976-80, as a squad leader, platoon sergeant and acting platoon leader. His most proud moment was being twice named Old Guard NonCommissioned Officer of the Quarter and working the Omar Bradley funeral.

His least proud moment was applying for food stamps to feed his family of four children. The most fun was at the post Spirit of America party in the Field House - he actually had the microphone leading a Charlie Guard cheer.

His most sad moments were as the NCOIC when a child was buried in the "boneyard." Most embarrassing moment was failing (one time only) the M-60 disassembly/assembly task at AP Hill.

His devoted, yet now late Army wife, Catherine, also served - rearing their young but large family in the cramped quarters of Tencza Terrace. No regrets, only sadness that I cannot do it again.

ALBERT G. BERTACCINI, born Sept. 27, 1920, Warren, OH. Graduated University of Pittsburgh, 1948 with BS in civil engineering. Commissioned: ROTC, April 25, 1942, Pittsburgh, PA; rank of 2nd lieutenant, US Army.

Assignment: Coast Artillery School (Sea Coast), Fort Monroe, VA. He was assigned to Btry. F, 55th S.C. (CA) FECOM, Dec. 21, 1942, Oahu,

HI. Tank Unit Commander, 263 CA, November 1943-April 1944; platoon leader, Anti-Tank Co., 3rd Inf. Regt., July 1944-March, 1945; adjutant, 4th Prov. Bn., 3rd Inf. Regt., March 1945-July 1945; commander, Anti-Tank Co., July 1945-February 1946; separation from service in June 1946.

Active duty, B Btry. 63rd AAAGN Bn., Fort Bliss, TX, November 1950; A Btry. and 63rd Btry. commander, August 1951-July 1952; battery commander, D Battery, 63AAA, January 1953-June 1954; staff officer, HQ Battery, June 1954-August 1954; Guided Missile Staff Officer School, Class #19, Fort Bliss, TX, August 1954-July 1955; Director of Maintenance & Director of Television School, Artillery-Radar System, 33 Nike Missile System, Fort Bliss, TX, July 1955-June 1960; Supply Office, Turkey-Chorlu, June 1960-June 1961; Contractor School (graduated), Pershing Missile System, July 1961-December 1961; Adj. 1st Msl. Bn., SETAF and Adj. Vicenza Military post, Vicenza, Italy, December 1961-1965; military advisor, National Guard Units, Pittsburgh, PA, January 1965-.

Retired Oct. 31, 1968, Camp Indiantown Gap, Annville, PA with the rank Lieutenant Colonel, USA (Ret.). His awards include the Army Commendation Medal, National Defense Service Medal w/OLC, Armed Forces Reserve Medal w/20 Year Device, WWII Victory Medal, American Theater Medal, Asiatic-Pacific Medal, EAME Medal, Army Occupation Medal (Germany) and three Overseas Bars.

Married Julia Laurenzi and they live in Old Forge, PA. They have two children, daughter Ruona M. and son Bruce J.

Major Employers: US Steel Co., 1948-49; Rust Engineering Co., 1949-50; City of Pittsburgh (traffic engineer), 1968-71; DeKald County, Atlanta, GA (roads and drainage maintenance supervisor), 1971-84. Retired in 1984. Old Forge Borough, PA, Public Relations Supervisor, retired in 1998.

He is a life member of American Legion, VFW and 3rd Inf. Regt. Old Guard Assn.; member of American Society of Civil Engineers and Old Forge Lions Club.

Hobbies are post card collecting, trains, covered bridges, animals, celebrities, Army (war equipment), Titanic memorabilia and history. Interests are CSS *Hunley* (Civil War submarine), history of the Titanic disaster and burial locations of Titanic victims.

JOHN E. BINKLEY III, born in Winston-Salem, NC on March 12, 1948. Enlisted in the US Army in September 1967, was commissioned February 1970, and retired a colonel of aviation in June 1995. Assignments included an enlisted tour with the 82nd Abn. Div. and commissioned service as the Salute Gun Platoon leader, 1970-1971, and executive officer, A Co., Third Infantry (TOG) 1972-1974.

Later assignments include tours with the 101st Abn. Div., 3rd Inf. Div., assignments as company commander, C Co. (1984-86) and later (1988-90) the executive officer and deputy commander, Task Force 160. He also served on the Army Staff (1986-88) as the Special Operations Aviation Staff Officer and from 1993-95 as deputy director, Air Wing, US Department of State. He served in combat with B Co., 2/327 Inf., 101st Abn. Div. in Vietnam and several campaigns with special operations forces.

His military education included Airborne, Ranger, Special Forces Officer Qualification, Rotary and Fixed Wing flight training, and Command & General Staff College. He has a BS in management from Columbus College and a masters of systems analysis from the University of Southern California.

He met and married Victoria Haynie in 1983. Upon retirement from the Army they relocated to Alexandria, VA where he pursued a career in aviation business development.

RICHARD M. BIRD, born Feb. 11, 1938 in Savannah, GA and reared in Middle Village, New York City. He enlisted in the Army in 1956. After basic training at Fort Carson, CO, he served in The Old Guard while stationed at Fort McNair in Washington DC.

After his honorable discharge in 1958, he began his career in advertising at the JC Penney Co. in New York City. Shortly thereafter, he started his own graphic design studio, and has owned his own business throughout his career.

He married Joy Loruenser in 1962 and has two children, Linda and Rich, six grandchildren and two step-grandchildren.

In 1989 he married Rita McGonagle, and they make their home in Manhattan. Richard is semi-retired and a consultant at Bird Advertising, Design & Fine Printers in New York City. He enjoys his family, travel, painting, and vacationing at his and Rita's second home in Merritt Island, FL.

JOHN C. BOYLE JR., Looking back he has many fond memories of being part of something that he can now look back and feel as if it were a lifetime ago. At this point of his life he prefers to think of the funny times.

Co. C was at the end of the street and the driveway ran along his barracks. Needless to say they needed a laugh, so one day they carried Michael Sullivan out on his bunk through the side door and left him fast asleep outside in the driveway. Hope Michael doesn't get angry when he reads this! John also thinks of another time on a color team at the tomb of the Unknown Soldier. He was a guard with Sergeant Pete Ganio and PFC Jimmy Floyd waiting for a Foreign Dignitary to arrive when a Bird Colonel walked up to them and said "You have got the flag on upside down."

John has been married for 31 years to Patricia and they have two children, Karen and John III.

JAMES R. BRADLEY, served three tours. First as a buck sergeant with B Co., 1973-76 as a line soldier (Infantry) doing jobs (Marching Platoon, Firing Party and Casket Team) in the bone yard. Then he came back to HHC as a specialist sixth class from 1978-81 as the Medical Section Sergeant/NCOIC having set up the Aid Station.

He served again in 1988-90 as the battalion medical officer and physician assistant. Any soldier that went on sick call will most probably have met him.

In 1992 he retired as a chief warrant officer 3 (physician assistant) after a 23 year career.

He now lives in New Port Richey, FL with his wife of 30+ years. They met while he was on his first tour and both did the Jitterbug routine in Spirit of America show.

Chief Bradley is a life member of The Old Guard Association and may be contacted at james.r.bradley@us.army.mil

ROBERT C. BUCKINGHAM, born in Wilkes-Barre, PA, June 21, 1927. Enlisted in ASTRP on his 17th birthday. After one year of college at CCNY and Syracuse, he completed infantry basic training at Camp Wheeler, Macon, GA.

Joined Old Guard, 3rd Bn., in Ludwigsburg, Germany in January 1946. Moved to Berlin on April 1, 1946. Assignments: Battalion Operations Sergeant then sergeant major until 3rd Inf. colors were sent to Fort Myers on Nov. 1, 1946. Discharged as tech sergeant Jan. 3, 1947.

Graduated from Wesleyan University and University of Rochester, School of Medicine with MD. Interned Harrisburg, PA Polyclinic Hospital and practiced family medicine in Harrisburg for 40 years. Then worked as National Quality Monitoring Contract Medical Director for TRICARE at KEPRO for nine years.

Married Carol Whiting in 1948. She died in 1986. Married Thelma Hoyer in 1988. They have eight children, 13 grandchildren and are expecting first great-grandchild. Retired in 2003 and reside in Hershey, PA.

MELTON E. CANNON, enlisted in the Army on June 19, 1979. He arrived at Fort Benning for One Station Unit Training (OSUT) and received the military occupational skill (MOS) 11B10.

1SG Cannon served with the 3rd US Inf. (TOG) from December 1991 to August 1997. 1SG Cannon served in numerous leadership positions: 6-man casket team, 8-man casket team, Platoon Sergeant First Marching Platoon, Platoon Sergeant First Presidential Marching Platoon (Joint Service), and Drill Master of US Army Drill Team.

While serving with the Casket Platoon, 1SG Cannon and his team were selected to perform Gen. Robinson and Ridgeway's funerals. In recognition of his superior performance, 1SG Cannon was selected to serve as the drill master of US Army Drill Team.

1SG Cannon will retire on Oct. 1, 2004 after serving as a first sergeant for a BCT unit at Fort Jackson, SC and pursue a teacher's certification or JROTC Army Instructor position in the Columbia, SC area.

DONALD B. CARMICHAEL, born in Los Angeles, CA on Feb. 16, 1925 and enlisted in the Army in February 1943. He was a combat soldier in three wars rising from private to colonel.

Landing on Omaha Beach he joined the 113th Cavalry and participated in five campaigns from St. Lo, France to the Elbe River, finishing war as a buck sergeant, tank commander. He commanded a tank platoon, rifle platoon and company in the Korean War as part of the 65th Inf., 3rd Div. in 1952-3 Served in Vietnam as a LNO, Capitol ROK Div. and later as deputy task force commander of Task Force South (two US airborne battalions and the 23rd ARVN Div., 1967-8.

Decorations include Silver Star, Legion of Merit, Bronze Star w/V and two clusters, Purple Heart, Air Medal w/2 clusters, Commendation Medal, ETO w/5 stars, Korean Medal w/3 stars, Vietnam Medal w/3 stars, Korean InHun (bravery), Vietnam Gallantry Cross, Belgium Croix de Guerre, Syngman Rhee Presidential, CIB w/star, Parachute Wings, Ranger Tab and many unit/service awards.

Joined The Old Guard as a first lieutenant in December 1954 and left for Fort Benning in December 1956.

Graduated from UCLA in June 1951 and received a regular army commission as a distinguished graduate; sought a masters degree at SFState prior to going to Vietnam. Post service as an insurance manager and taught courses at Pima Community College in Tucson, AZ.

Married to Sally Jeanne and they have five children and many grandchildren. They live in Crestview, FL in the Panhandle.

He has written three books - one each on the wars; speaks to religious, community and military organizations; and is active in church functions.

RALPH CASADO JR., born Aug. 25, 1935 in Brooklyn, NY. Enlisted in the US Army in February 1955. Assigned to The Old Guard Headquarters Company, Commanding Officer Col. Fielder P. Greer.

As a draftsman he laid out the parade ground for regimental reviews, redesigned the regimental commander's office and regimental briefing room and was in the platoon sent to the Washington Premiere of Audie Murphy's biographical film, *"To Hell and Back."*

Upon discharge in 1957 he was awarded The Good Conduct Medal and in 1998 received a "Cold War Certificate of Recognition" from Secretary of Defense William S. Cohen. He is a charter member of The National WWII Memorial and The American Air Museum in Britain.

He graduated from Brooklyn College with a BFA Degree and worked in the advertising profession. When he retired from his own agency, Promotion Solutions, in 1995, he became an adjunct professor at The New York City College of Technology where he currently teaches computer assisted advertising design.

He has three grown children: Robert, Richard and Lisa, and resides in Millerton, NY with the love of his life, Phyllis.

JOHN JAMES CHANDLER SR., born Dec. 12, 1932, Graham, NC. Enlisted May 3, 1949, Danville, VA and was discharged Nov. 15, 1952, with the rank of staff sergeant – no noteworthy medals.

After six months of basic training at Fort Knox, KY, he was transferred to The Old Guard. He had no idea that it was a spit and polish and a ceremonial unit for the President. He is very proud to have been a part of that elite unit. The Third Infantry has a long history. It has fought in all wars.

Some of his Old Guard duties include escort for funerals at Arlington Cemetery; Guard on the Tomb of the Unknown; Guard at the Pentagon during the Korean War; Guard at parades, civilian shows and for Joint Chief of Staff on the Post; and met high ranking dignitaries at airport.

Received orders to go to Korea. After he arrived in California, he was informed that he did not have enough time left in the service, so they changed his orders and sent him to Fort Ord, CA to train recruits. He thinks all young men should spend a couple of years in the service. It helps to make a man of them. It did him.

The best thing about his service was meeting young men from all over the country. Some became just like brothers. The downfall is that you lose a lot of good friends during the war. His time in The Old Guard was a good experience and he would not change any of it.

After discharge, he attended night school with on-the-job training for five years (apprenticeship as a plumber) under the GI Bill. He worked several years as a journeyman plumber for a few contractors. After four years, he took the examination to obtain a master plumbing and heating license and received all Northern Virginia and State licenses. In 1969, he started his own business and continued in that business until retirement in 1987 at Arlington, VA.

Met his wife, Grace Ann Goldblatt, and mother of their children at the North Post NCO Club She was the best mother in the world and died July 13, 2000 after 48 years of marriage. She is waiting for him in Arlington Cemetery.

His four children are Victoria Chandler Elmore, John James Chandler Jr., Valerie Chandler Painter and Vanessa Lynn Chandler. Grandchildren are Misty Beth White, Nathan Daniel Elmore, John James Chandler III, Kelcie Julianne Chandler, Kacie Lynn Painter, Kellen David Painter, Keven Lee Painter and Kory Lee Painter.

He's a Christian and studies and compares all religions and also the Civil War. He lives in Culpeper, VA.

DENNIS W. CHOPLIN, entered the Army on Feb. 19, 1959 at Winston-Salem, NC. After basic at Fort Jackson and AIT at Fort Hood, TX, he was assigned to the 1st Battle Group, 3rd Inf. Regt. Tank Platoon of HHC.

When the unit was re-organized as the 1st Bn., 3rd Inf., the tanks were turned in and a Recon Platoon formed. He elected to remain with the unit for 4-1/2 years, leaving the Army as a SSG E-6.

He re-enlisted a year later going straight to Korea. From there to Fort Riley, KS to Kaiserslautern, Germany. From Germany he volunteered for Vietnam and was assigned to the 4th Bn., 3rd Inf. (The Old Guard). From there back to Fort Riley, KS to Bamberg, Germany and on to Duluth IAP, Minnesota where he was the only personnel assigned to the Air Base of multi-service.

During his service he was awarded the Defense Meritorious Service Medal, two Bronze Stars, three Army Commendation Medals, National Defense Medal, Vietnam Campaign Medal w/2 stars, Vietnam Service Medal, Vietnam Cross of Gallantry w/palm and the Good Conduct Medal Seventh Award. He also earned the Jungle Expert Patch while TDY to Panama Canal Zone.

After retiring in 1980 he earned his associate of applied science degree and worked at the University of Nebraska for the next 15 years. He retired again in September 2002 and became a triple dipper.

ROBERT H. CLARK, born Jan. 19, 1932, in Georgia. A distinguished member of the regiment, he served two tours with The Old Guard: first as executive officer from 1965 to 1967 and secondly as commander from June 18, 1974 to Jan. 30, 1976. Served as MDW Ceremonies Officer following assignment as Old Guard XO.

Earlier army service included duty with the 82nd Abn. Div. and the 25th Inf. Div., Hawaii. Other overseas duty included two tours in Vietnam: 1963-64 and 1969-70, the latter as commander, 4th Bn., 12th Inf. Prior to assuming command of The Old Guard, he served as senior US observer in Middle East (Palestine) and charged with monitoring cease-fire of Yom Kippur War. Graduate of various army schools including CGSC and the Army War College.

Decorations and awards include: DSM, LOM w/OLC, DFC, BSM w/2 OLCs, AM, JSCM, Combat Infantryman Badge and Senior Parachutist Badge. Retired from active duty in 1979 following assignment as executive officer to the DCSPER, HQ, DA.

Still married to Nancy and they have one son and one granddaughter.

DONALD G. COFSKY, was born in Brooklyn, NY in 1934, graduated from Colby Academy, attended Brooklyn College for degree in business administration and sales management.

He was inducted into US Army Jan. 12, 1955, and after basic and AIT at Fort Dix, NJ, was assigned to 3rd US Inf. Regt. - The Old Guard. Served in casket team and firing team positions, and was appointed assistant squad leader and firing team leader. He subsequently attained rank of sergeant and was appointed squad leader.

Don left military service Jan. 11, 1957, entered industry as a sales representative for a major soft goods manufacturer Over the next 10 years he held various sales and sales management positions in the ladies ready-to-wear and associated industries. Moved to Rochester, NY in 1968 as a regional sales manager for a major manufacturer, and changed career path in 1971 and entered the personal lines insurance industry, eventually rising to assistant brokerage manager. This position led to contact with business owners in the real estate and home building business, and in 1984 Don established his own residential marketing company, Market Trends, a full service marketing organization for regional and national residential home builders. In 1985 he was selected by the National Association of Home Builders as the first Upstate New York "Member of the Institute of Residential Marketing," a distinction he maintained until he retired from the industry in 1992. Over the years he has served as a motivational speaker for the National Association of Home Builders, appearing at national conventions and local Home Builder Association meetings and seminars around the country.

Don became actively involved with various Veteran's organizations in 1990, and served two terms as president of the Monroe County (Rochester) Chapter of the Korean War Veterans Association, 1992-93 and 1998-99. In 1994 he was elected president of the Department of New York, KWVA, and served two one-year terms. He concurrently served as national chairman of the KWVA "TELL AMERICA" Committee, a public awareness program, 1992-95.

Drawing upon his experience with The Old Guard, in 1993 Don organized the KWVA Monroe County Color Guard. He still serves as commander of the unit which appears at regional parades and public events and serves as Honor Guard for fallen comrades.

In 1996, Don was selected by the Monroe County Executive to serve on the County Veteran's Service Agency Board of Directors, a position held until 2001.

In 1995, Don formed and chaired the committee to design, build and dedicate the Greater Rochester Korean War Veterans Memorial, and in 2000, formed and chaired the committee to design, build and oversee the Rochester Korean War Commemorative Park, as a "Commemorative Partner" with the Department of Defense's 50th Anniversary of the Korean War Committee. In 2001 the park, located on NY Route 441, was designated by the New York State Legislature as the "Korean War Memorial Highway," a legislative effort initiated and championed by Don.

Don enrolled as a charter member of The Old Guard Association (TOGA) in 1998, and is a life member. He volunteered as the organization's first newsletter editor, a post he retired from at the end of 2004. He was elected to serve on the TOGA Board of Directors for two 2-year terms in 2000 and 2002, and was named president of the Association, 2002 and 2003. He was nominated as a Distinguished Member of The Regiment on Veteran's Day, Nov. 11, 1999, a distinction he describes as "one of the three greatest moments in my life, after my marriage to Sharon and the birth of my son, Jeff." Don currently serves as the Association's Liaison to The Regiment.

Don also served as the executive editor of the premier edition of The Old *Guard/Old Guard Association History and Yearbook,* which is in production for publication in early 2005.

Don and Sharon reside in Fairport, NY, and are the proud grandparents of Nicole and Jacob, who live in Houston with their parents, Jeff and Adriana.

GERALD W. COMER, born in Bartow, FL, July 7, 1926. (SGM Ret.). Drafted into Army in November 1944, trained at Camp Blanding, FL. Sent to Europe during WWII to the 78th Inf. Div. After war attended SoGA College.

Enlisted June 1949 and assigned to Old Guard, Fort Myer, VA until 1952. Other Assignments: Korea 25th Inf. Div.; Fort Hood, TX, 1953; ROTC Duty, Dallas, TX (1954-60); 26th Inf. Regt., 8th Inf. Div., Germany, where he moved from Infantry to AG; Personnel Admin.; Fort Benning, GA, 1963; Korea, 1967 and returned to Fort Benning, GA; Vietnam in 1972 and returned to Fort Benning where he retired in November 1973.

Developed the Learning Centers, Fort Benning, GA. Met and married WAC, Gladys Pritchard in 1950. They have two children, Linda and Larry, five grandchildren and one great-granddaughter. They now live in Columbus, GA.

His awards include the Combat Infantry Badge, Legion of Merit. Bronze Star w/V, Bronze Star(A), Meritorious Service Medal and ARCOM.

LLOYD NEALE COSBY, born Feb. 3, 1933. He entered the US Army in September 1955 and joined The Old Guard May 30, 1958. He served as

platoon leader of the guards at the Tomb of the Unknown Soldier in Arlington National Cemetery.

Attended the University of Kentucky and master's degree from George Washington University. (See Special Story section for his memorable experience).

STEPHEN COTTON, Specialist 4th Class (E-4) with the 1st Bn., 3rd US Inf. "The Old Guard" (Bravo and HHC companies) from 1981-84 (ETS). Member of 1st Casket Team and Battalion Commander's driver.

Stephen was born in St. Petersburg, FL in 1962. Enlisted in USA as E-3 on Aug. 28, 1980 at age 17, from Dixie Hollins High School. Graduated from JROTC, entered basic training/AIT with Bravo Co., 7th Bn., 1st Inf. Tng. Bde., Fort Benning, GA where he was Platoon Guide.

Recipient of ARCOM, GCM, Driver/Mechanic Badge, Expert Badge, Army Service Ribbon, Cold War Certificate of Recognition. White House Duty Access with congruent security clearance.

An author, poet, historian, and philosopher. Distinguished Member International Society of Poets, inducted into Poet's Hall of Fame. Life member Old Guard Association. Member of noncommissioned Officers Association. Keeper of the memory of His Grace the 10th Duke of Atoll, George Ian Murray. Continues in active civilian federal service.

RICHARD L. CRAIG, born on June 13, 1929 in Greenville, PA. He was drafted on March 13, 1951 from Mercer County, PA and inducted in Erie, PA. He was processed at Fort Meade, MD and selected for assignment to The Old Guard, the Presidential Honor Guard, 3rd Inf. Regt. at Fort Myer, VA. He received basic training at Fort Myer with field training at both Fort Meade and Fort A.P. Hill. After basic graduation in late July, he was assigned to Line Co. B; serving his entire two years duty time there. He was discharged on March 12, 1953.

He participated in quite a few historical events: 1) MacArthur's Retirement Parade in May 1951. His speech in front of Congress provided the famous line, "Old soldiers never die - they just fade away." 2) Standing in review for King George VI and Queen Elizabeth of England on their visit to the US in September 1951. 3) Standing in review for President of Chile in October 1951. 4) President Eisenhower's Inaugural Parade on Jan. 20, 1953. Their 366 man unit marched from the front of the US Capital Building down Pennsylvania Avenue to the White House.

Richard also played on a post baseball team in the summer of 1952, along with four major league players: John Antonelli, Bob Purkey, Danny O'Connell and Sam Calderone. Richard was the only nonprofessional player on the team.

Upon his discharge in March 1953, he worked in wholesale food sales until March 1962. He then became a career agent with the Mutual Life of New York (MONY) for 30 years. He also established a brokerage business and continued that while phasing out when his son joined him. His son, Christopher, continues to run the business today.

Richard and Sheila have been married 44 years and in addition to son Christopher, also have two daughters, Erna and Kelly.

RICHARD E. CROSS, Colonel, USA, Ret., assigned to the 3rd Inf. (The Old Guard) from 1961-64. He was called to active duty in the US Army as a private in 1943, while attending "The Citadel," military college of South Carolina. He served on active duty from 1943-73.

During his service he was promoted to the rank of sergeant and later to colonel. In the summer of 1944 he was assigned to the Infantry Officer Candidate School at Fort Benning, GA and was commissioned a 2nd lieutenant on Jan. 9, 1945. The US Army Officer Candidate School inducted him into the Hall of Fame on April 6, 2001.

His military career included overseas assignments to the Philippines, Japan, Germany, Alaska, Burma, Vietnam, Panama and Bolivia. He served with the 11th Abn. Div., 82nd Abn. Div., 1st Inf. Div. (The Big Red One), and the 3rd US Infantry (The Old Guard).

Other assignments included the Army General Staff and Department of Defense Staff, Washington, DC. With the exception of a two month interim assignment in the Summer of 1963 as Assistant Chief of Staff, Military District of Washington, his assignments with the 3rd Inf. were that of S-4, Executive Officer, Deputy Commander and Commander of the Regiment (53rd).

During the latter part of 1963 The Old Guard and other units of the Armed Forces within the District of Washington served with distinction in the funeral of the late President John F. Kennedy. In February 1964 he was assigned to the Department of Defense, Washington, DC.

Col. Cross is married to the former Lilo M. Godfrey and has three sons: Richard E. Cross Jr., Stephen Denton Cross and Thomas Joseph Cross.

ROY S. CUMMINGS, born Nov. 5, 1941. He enlisted Jan. 10, 1961. Basic training was at Fort Dix, NJ and AIT at Fort Benning, GA.

Sent to Korea for 15 months. Reported to A Co., The Old Guard in September 1962 at Fort McNair. Main job was casket team at Arlington and Culpeper, VA, Cordons, Sunday parades at the Washington Monument, M.L.K.'s "I Have a Dream" speech-ready to go with live ammo to protect the White House in case of riots. Pres.

Kennedy's Funeral was his saddest moment. Lt. Samuel R. Bird was their platoon leader, Sgt. Major William Mashburn was their first sergeant.

Discharged Jan. 9, 1964 as SGT E-5. Received the Good Conduct Medal.

First wife was Joyce (Matusky) and they had children: Matthew, Victoria and Stephen. She passed away in 1987. He married Karen (Stephens), child-Michael. Grandchildren are David, Alexis and Laticia. They reside in south Florida. Hobbies are golf, drag racing and bike riding. Truck repair, leasing, logistics is what he does now.

WILLIAM R. "BILL" DAKNIS, born Aug. 27, 1934, Philadelphia, PA. Commissioned 2nd lieutenant, Infantry, 1957 from Drexel University, Philadelphia, PA. Assignments include duties at Forts Benning, Polk, Lewis, Belvoir, Dix, Ord, Monroe, Schofield Banks and the Ohio State University.

Overseas assignments were Commander, 1st Bn., 8th Inf., Korea; inspector general, 193rd Inf. Bde., Republic of Panama; 36th Ranger Bn. advisor, staff officer, 11th Inf. Bde., Republic of Vietnam and staff officer w/Commander, Amphibious Group Two aboard USS *Mount Whitney*.

Retired in September 1986 as colonel. Awarded Silver Star, Bronze Star w/OLC, Legion of Merit w/OLC, Air Medal w/OLC, Meritorious Service Medal and Purple Heart. Selected as the recipient of the first Col. Leo A. Codd Award as the most outstanding ROTC Instructor in the US in 1972.

Upon retirement from the Army he was employed as senior evaluator for Vinnell Corp. serving in Riyahd, Saudia Arabia, 1987-89. Was director of nonresident instruction at the Army Staff Management College in 1991.

Married to the former Elizabeth "Bettie" McLaughlin since September 1957. Blessed with four sons: Bill, Bob, Steve and Mike. Additionally blessed with four daughters-in law: Mishie, Anne, Wendy and Jennifer. They have 13 wonderful grandchildren: Shaun, Ryan, Mike, Marc, Zach, Kristen, Meg, Jenny, Will, Chase, Kate, Sarah and Griffin.

Bettie and Bill currently reside at Watergate at Landmark in Alexandria, VA.

TIM DAVIS, Sergeant, 2nd Platoon Co. B, 3rd US Inf. (The Old Guard). Tim was a good friend and he saved my life while on a training exercise at Fort Carson, CO while they were assigned to the 1st of the 8th Inf. Bn. *Submitted by Carl L. Hardy.*

GEORGE "JERRY" DEADY, born in New York City, NY. He joined the Army and after initial training at Fort Dix, NJ, was assigned to the 1st Inf. Div., Fort Riley, KS.

From 1965-71, his assignments included four years in Vietnam with the 22nd ARVN and 25th Inf. Divs., and two tours as an instructor at the Infantry School, Fort Benning, GA. In 1971 SSG Deady was assigned to H Co, TOG, where he served under 1SG "Jack" Daniels and really became an NCO, then E Co, Honor Guard.

Other assignments include the Berlin Brigade, 1SG, Infantry Training Center, and NCO Academy Company Commander, Fort Benning, GA. In 1994 he retired with 30 years as sergeant major of the NATO HQ in Izmir, Turkey, where he currently lives and works as an US Army civilian. He married his wife, Mukerrern "Muki" Aldemir, in 1998.

WILLIAM ALVA DEFREITAS, born May 23, 1943, St. George's Grenada, West Indies. Enlisted July 21, 1965, at White Hall Street, New York, NY. Basic training and AIT at Fort Jackson, SC. Assigned to Co. E Honor Guard 1st Bn., 3rd Inf. at Arlington National Cemetery, Fort Myers, VA. Position: Firing Party.

Discharged July 21, 1967 with the rank SP4. His awards include the National Defense and Good Conduct.

Accomplishment: Director of School Program for United Way Sports and Arts, New York.

Civilian Activities: NCAA, FIFA, High School Soccer Referee, High School and University Coach. Currently teacher of PE in a high school in Brooklyn, NY. His hobby is collecting foreign notes (money).

Married Nov. 25, 1972 to Lesley G. Hoyes. They have three children: William Alva, Desmond Ivan and Nigel Ian; grandchildren are Octavia, Xavia, Shane, Nadia and Zaria.

FRANCIS GAVIN DOLAN, born in Waltham, MA on Sept. 22, 1930. Graduated from St. Charles High School, Waltham MA. On Feb. 29, 1952 he began US Army training at Fort Indiantown Gap, PA, was later transferred to Fort Myer, Arlington, VA, The "Old Guard." First assignment was with A Co. and later transferred to HQ Co., 1st. Bn. 3rd Inf. Regt.

Service was during the Korean War; however, he did not serve in Korea and was discharged on Feb. 28, 1954 with a rank of corporal. Awards include the National Defense Service Medal, Good Conduct Medal, Sharpshooter Medal, Korean Conflict and later Cold War Certificate of Recognition.

A very memorable experience among many was when HQ Co. planned the Inaugural Day Parade for President Dwight D. Eisenhower. It was the first Inaugural Day parade for any President to be televised nationally. The 3rd Inf. Regt. received a commendation from the US Army Chief of Staff, J. Lawton Collins.

Civilian activity included working for a National Manufacturer in Watertown, MA, and Syracuse, NY and was later transferred to San Antonio, TX and eventually became vice president of the company. Married Ann (Pewtherer) Dolan in 1952 in Waltham, MA. Ann died in 1995. Met Barbara (Austin) Dolan who lost her husband and they were married in San Antonio, TX in 1997. He and Ann had five children. Barbara has two children. There are nine grandchildren. Francis and Barbara live in San Antonio (Fair Oaks Ranch), TX.

It was a pleasure and honor to serve in the "Old Guard." What a wonderful experience to be part of the great tradition of military vets and servicemen who have contributed so much to the freedom of the United States.

JOSEPH R. DONATO, born July 23, 1932. Enlisted at Philadelphia, PA and took basic training at Fort Dix, NJ, Co. D 1st Bn., 60th Inf. Regt., 9th Inf. Div.

On Dec. 15, 1949 he was shipped to Fort Myer, VA, joined The Old Guard Dec. 16, 1949 and assigned to Ceremonial Co. (later changed to Co. A), 3rd Inf. Regt. Assigned to regular company duties: firing parties, casket bearers, gun detail, Color Guard, drill platoon, guides, full honor funerals, etc. Assigned to tomb guard April 15, 1950 to April 1951, sentinel duties for one year.

Discharged Dec. 16, 1952 with rank of corporal (old buck sergeant rating in those days). Prior to enlisting he worked as a stunt man in Hollywood in the old Roy Rogers movies, rode bulls, and bareback broncs in rodeo. Later he entered movie business in a variety of jobs until he became executive producer and produced over 100 films from 1977 to present. He lives in Sherman Oaks, CA and is still employed as same.

RONALD A. DOOLEY, Sergeant, Co. A, 1958-59, born Oct. 9, 1937, Riverside, CA. Entered service June 1, 1955 at Fort Ord, CA. Served in Fort Ord, CA; Fort Carson, CO; Fort Mead, MD; 24th Inf. Div. in Korea; Old Ironsides at Fort Polk, LA; 3rd Inf., The Old Guard at Fort McNair, WA, DC. Left service June 30, 1959, Fort McNair, WA, DC.

Ronald and David Handley think they had the absolute best top kick any company could ask for – CSM James Trepoy, who now lives in Salina, KS and still keeps in touch with many of his troops from those days. Through CSM Trepoy they had a sharp company and won many parade top awards. Their most memorable and taxing job was that of the Double Interment Ceremony in May 1958 which lasted three days (but for them it was much longer after many days of rehearsal).

One funny happening when someone stuffed squirrels in the cannon and when the cannon was fired for Retreat it was quite a mess. The noncoms and officers living on post didn't see the humor. The post MP Detachment was blamed, but they never confessed.

Married in 1961 to Pattie and they have three children, four grandchildren and three great-grandchildren. Their oldest grandson, Chris Hille, is in the Marine Reserves in San Antonio, TX and soon to deploy to Iraq.

Ronald served 43 years in law enforcement/criminal justice, retired Nov. 1, 2002 and now lives in Fairfax Station, VA. Currently a Mason in Scottish Rite, National Sojourners, Heroes of 76, and B&P Order of Elks, Order of Eastern Star, Fairfax County Retired Police Association, and a life member Old Guard Association.

EDISON L. DOWDY SR., born Oct. 11, 1917, in Newport, VA.

Mr. Dowdy was drafted and inducted into the Army, 3rd Inf. Regt. on Nov. 23, 1943. At that time, he was married to Lyda McCoy of McCoy, VA. They had one son, Eddie. He also has two grandchildren and three great-grandchildren.

He was a PFC and was stationed at Fort Benning, GA and was later transferred to Germany.

Mr. Dowdy was a graduate watchmaker and jeweler. He and his late wife founded Dowdy's Jewelers in Pearisburg, VA and that business continued for 44 years. *Submitted by his son, Eddie Dowdy.*

CHARLES A. DRAKE, was a casket bearer with Delta Co., 3rd US Inf., Fort Myer, where ETS'd in 1979. Those Tws were sure a nice uniform. Some of the best times of his young adult life were spent with those buddies in Delta Co.

He went on to college, but stayed in the Maryland Army National Guard, doing most of his time with Bravo Company, 20th Special Forces Group; 2003 found him still in after 26 years (now a MSG) with the Special Operations Detachment, Joint (SODJ). Boy, its been some trip. And it ain't over yet!

MARK RANDOLPH DUCRE' (SMITH), born May 6, 1955 at Fort Lawton Army Hospital, Seattle, WA. Enlisted May 22, 1973, HHC, 3/160th Inf. (Mech) 40th Inf. Div., CALANG-36K20.

Enlisted in TOG June 3, 1974 and served in Charlie Guard, 1974-75-11B20, HHC Bn. Commo 75-77-36K20. No combat experience.

Discharged April 10, 1979 with the rank SP4. His awards include the Good Conduct and National Defense Medal. Accomplishments: served as Salute Guns Signal Liaison for G.O Funerals, National Ceremonies and Departure of President Gerald R. Ford at Andrews AFB during Carter Inauguration.

Married Lorraine Mae Long on June 27, 1998. They have three children: Kristal M. Duvallon; SSG Timothy M. Smith, 101st ABN; C/1LT Joshua Smith AFJROTC. Grandchildren are Emanuel, Matthew, Mark Anthony, Timothy, Alberto Jr., Patricia, Kristen, James, Keith, Roxanne and Angelica.

Currently living in Las Vegas, NV and working fulltime as a sales associate for Home Depot. Hobby is gourmet cooking.

STEVEN M. DUNFEE, born Harrisburg, PA, June 10, 1958. Enlisted in 1975, Tucson, AZ. Basic training was at Fort Ord, CA. Assigned H Co. 3rd US Inf., The Old Guard, Fort Myer, VA until active duty ETS in 1978. Attained the rank of Spec 5. The Good Conduct and Army Commendation Medal were awarded while assigned to The Old Guard. Reserve ETS November 1981.

He met and married Laura Blocher while in The Old Guard in 1977 and had first child, Grace, while stationed at Fort Myer, VA. She was born at Walter Reed Army Medical Center in 1978. After active duty service he entered the Tucson Police Academy graduating in 1980. His second child April was born in 1980 while he was in the police academy. His police career spanned 20 years with various assignments including street patrol, the Fraud Division, the Narcotics Division, and finally the Street Gang Unit. He finished his career as a detective with numerous awards and citations.

He is currently a senior vice president and national sales director with Primerica Financial Services, a subsidiary of Citigroup. He is a recognized researcher of American military history emphasizing the Federal American Civil Soldier and Civil War Infantry Tactics. He has consulted for several museums and historical foundations and has worked on several historical preservation projects, movie productions, and documentaries.

He currently serves as the lieutenant colonel of the Army of the Pacific, a nationally well-respected living history organization and publishes a catalog specializing in personal artifacts of the American Civil War period.

Steven and his wife reside in Tucson, AZ, and he is a life member of The Old Guard Association.

JOSEPH A. DYER, born Sept. 25, 1921, in Dahlonega, GA. Enlisted in the Army while attending North Georgia College in 1942. He attended Clemson College and the Army Specialized Program.

He graduated from Infantry OCS program in June 1944 and was ordered to duty with the 42nd Div. at Camp Gruber, OK. While there, after eating too many field ration candy bars, he was diagnosed with diabetes and placed on six months of limited duty. He did not go to Germany with the 42nd ID.

With his limited duty he was assigned to the Infantry Replacement Training Center at Camp Joseph T. Robinson, AR. If that was light duty, he was glad it wasn't heavy duty! From December 1944 until June 1946, he helped train four cycles of basic trainees. This required 15-hour days, six days a week.

He guesses it was the good exercise that helped get his sugar under control, and he didn't get kicked out of the Army. He specialized with the mortar weapon and later in his Army service taught it with the Weapons Department at the Infantry School.

After the war he was transferred to the Corps of Engineers. After schooling at the Army Map Service he was assigned to the 660th Topographic Battalion in Panama, where they were mapping Colombia and Costa Rica. After several months in the jungle, he decided the engineers was not for him. He left the engineers to go on competitive tour for a Regular Army commission, After one year he was awarded a Regular Army commission in the infantry.

After his tour in Panama, he was assigned to the 3rd Old Guard Regiment at Fort McNair, Washington, D.C. He arrived there in June 1950, one week after the beginning of the Korean War.

In 1946 he married an Army nurse, Lucille Townsend, from Pennsylvania. She had served in Germany with a hospital in support of the 82nd Abn. Div. That marriage lasted 20 years, and together they reared three wonderful sons. One, Dr. Raymond B. Dyer, is one of the top radiologists in the country, and has written three volumes on the kidney and its functions. Joseph T. Dyer graduated from Georgia Tech with a master's degree and also from Harvard Business School. He has been part owner of several companies, and has done all this while confined to a wheelchair. He became a paraplegic as a result of an accident in Germany at the age of 16 which occurred while driving a Volkswagon, known as Hitler's Curse on Humanity. His youngest son, Lynn F. Dyer, is an automotive technician, who is ASE qualified in many fields. He also has two wonderful grandsons: Chris Dyer, 18, who is an Eagle Scout and will enter college this fall, and Richard Dyer, 13, who shows great promise in the liberal arts field.

In 1970, he married Shirley Bergeron, a Department of the Army civilian who worked in the Pentagon. She brought to our family a daughter, Dana Dyer Schalk, who graduated from the College of William and Mary and is employed on the corporate staff of a major bank.

After spending 28 plus years in the Army, he retired in 1970 as a lieutenant colonel. During those years he had five tours in the Washington area, one on the Army General Staff, one on the Defense staff, and the last with the Army Inspector General. He served in five divisions with two tours with the 7th ID in Korea.

In 1991, Shirley and he bought their retirement home in beautiful Roanoke, VA. He is in fair health at 82, but there's not much that he can do anymore except take his pills and pay his bills. He thanks God for his wonderful family and the enjoyment and privilege of living in America. He hopes and prays that our country can start to realize what we possess and come together.

DOUG "TEX" EDINS, born in Abilene, TX, Sept. 15, 1942. At age 21 he became a Texas Highway Patrolman. While in the DPS Academy, his turn came to be drafted. The DPS secured a deferment for his four months in the Academy and for 12-months after his graduation as he served out his probation as a new Highway Patrol officer.

In 1965 he was drafted into the US Army. After basic and advanced training at Fort Polk, LA, he was assigned to The Old Guard and served in First Platoon of Co. E (Honor Guard) and as a member of the US Army Drill Team.

Discharged April 11, 1967 he returned to Texas and to his DPS job where he retired in 1993. Later returning to DPS in a civilian capacity, where he remains today.

His two years in The Old Guard remain some of his fondest memories.

THOMAS EVANKO, born in Bridgeport, CT on April 4, 1955. Enlisted in the Army in December 1977. Joined The Old Guard in November 1978 and was discharged from the Army in December 1980.

Awards include Basic Parachute Wings, Expert Driver Badge, Army Commendation Medal, Navy Achievement Medal, Army Good Conduct Medal, Navy Meritorious Service Medal, and various other medals and ribbons.

Retired from US Navy Reserves after 23 years of total service both Army and Navy as a Boatswain's Mate 1st Class. Old Guard Units - D Co., H Co. (Recon), also worked as the driver for Battalion Aid Station.

Best Memory - As member of the RECON Platoon Old Guard, which was attached to the 1st 75th Ranger Battalion in Panama Canal Zone during the hand off of the canal to the Country of Panama in 1979. The Old Guard football team, made up of members of the RECON Platoon beat the best of the Rangers in a tackle football game in front of the entire Ranger Battalion. This devastated the Rangers!

Today, he is married and has one daughter. He still lives in Connecticut and works in sales.

ROBERT A. FINTON, born Feb. 6, 1959, Baltimore, MD. Enlisted February 1979, active Nov. 2, 1979; basic, AIT, and airborne at Fort Benning, GA. Assigned D Co., 1st Bn., 3rd US Inf. (TOG). Arrived Fort Myer, VA in April 1980 and served with 2nd Plt. (Caskets) and HQ Plt. (Commo/NBC). Attended NCO School at Fort Knox, KY; promoted Sgt. (E-5) in December 1982. Served as a squad leader and New Man instructor.

Left active duty Nov. 1, 1983 and served with B Co., 20th Special Forces (MD National Guard) until February 1985.

Received BS degree in zoology, Towson University in 1987. Educator with the Maryland Science Center, 1988 to 2000. Currently the Education Coordinator for the Chesapeake Bay National Estuarine Research Reserve (a NOAA program).

Married fellow educator Jeannine Holdiman in May 1993. Son, James, born in September 1994 and daughter, Gillian, born in March 1997. Currently living in Joppatowne, MD.

JAMES C. FOSDYCK, born in Amarillo, TX. Sept. 28, 1948 to Clarence E. Fosdyck and Asta Mae Sullivan Fosdyck. James was inducted into the Army May 7, 1968. He completed basic combat training and advanced individual training (Infantry) at Fort Ord, CA. His other assignment was with the 3rd US Army Inf. Regt. (The Old Guard) at Fort Myer, VA, where, as a member of 1st Plt. B Co., he participated in the Presidential Inauguration of Richard M. Nixon and the funeral of former President Dwight D. Eisenhower. He was honorably discharged May 6, 1970 at the rank of Sergeant (E-5).

James married Un Hui Yi Sept. 8, 2002. He has two children from a previous marriage, a daughter-in-law, son-in-law and two grandchildren: Wesley J. Fosdyck, Melinda Dent Fosdyck and their son Alex, and Jeanne-Marie Fosdyck Siegel, Kevin H. Siegel and their daughter Kaia Mae.

James served for more than 33 distinguished years as a law enforcement officer with the Los Angeles County Sheriff's Department having been awarded that Department's Distinguished Service Medal, Meritorious Service Medal, Sheriffs Medal and Service Recognition Medal. He retired from law enforcement Jan. 20, 2004.

James, Un Hui and the other members of their family reside in Orange County, CA. His hobbies include traveling, gardening and family history research (he published his family's history, History of the Fosdick - Fosdyck Family in 1998).

JOHN PETER GANIO, born in Danbury, CT, May 30, 1941. He was the second of Maria and Dario Ganio's four children. He graduated from Western Connecticut State University in 1965. He married Bernadette Kernicky Ganio in 1966. They have one son, Peter Nicholas Ganio. He is currently the vice president of a printing company and was involved in banking for many years.

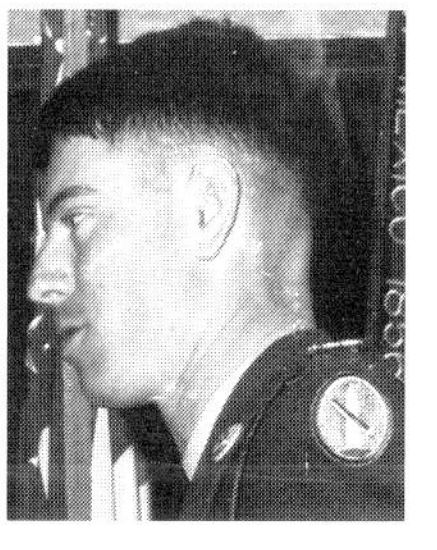

He was drafted into the Army in April 1967 and served until April 1969. He was assigned to The Old Guard and stationed at Fort Myer from September 1967 to April 1969. He was a sergeant in the C Co. 1st Plt. Color Team.

John was privileged to participate in President Eisenhower's funeral, Richard Nixon's inaugural and Bobby Kennedy's funeral. He was also involved in riot control in Washington DC following Martin Luther King's assassination. The greatest honor was serving at the funerals of those lost during the Vietnam War.

He was honored to serve in The Old Guard during a significant period of our country's history.

JOE L. GARCIA, born in Whittier, CA, July 20, 1947. Resides in Walnut, CA. Drafted in February 1968.

Basic and AIT at Fort Ord, CA. Arrived at Fort Myer in June 1968. Assigned to Bravo Co. 2nd Plt. (Caskets). Participated in President Nixon's Inaugural, funerals for General of the Army/President Dwight D. Eisenhower, Sen. Everett Dirksen, MG Keith Ware (Division Commander 1st Inf. Div. KIA Vietnam) and countless others.

Discharged in February 1970 with rank of SGT E-5. Earned National Defense Medal, Good Conduct Medal and EIB. A salute to all those he served with, wherever they may be.

Married in 1983 - one son, Andrew, born 1986 who is graduating from Loyola High School in Los Angeles and will be attending University of California Santa Barbara in fall of 2004. Joe worked for Pacific Telephone, AT&T and Lucent, retiring in 2000 after 28 years. Coached youth baseball 20 years. Joe's father is a Navy veteran of WWII.

RICHARD M. GARRETT, born in Barberton, OH and graduated from Norton High School. Served a tour of duty, 1948-51. Drafted in March 1951 and trained at Fort Myer Provincial, Co. H, Fort Myer, VA.

Selected to serve in A Co. 3rd Inf. Regt. At that time the Tomb Guard had vacancies. He went to the 3rd relief and served from October 1951 to March 1953. Walked post for one year, was promoted to corporal of the Guard for six months. Pictured in *Movie Star Magazine* September 1952 with Janet and Tony Curtis. He participated in the Inaugural Ceremony for President Eisenhower in 1953. Richard was discharged in March 1953.

Employed at Seiberling Tire and Rubber Co. In 1956 he married Justine Winebrenner, son Bradley was born in 1960 and daughter Beth in 1962. Brad married Marlene Mazey and grandson Bradley C. Garrett was born in 1991. Richard retired from Seiberling Tire after 30 years, then worked at Akron Recycle until final retirement. His hobbies are serving on the board of the Degenhart Glass Museum in Cambridge, OH; trap shooting and collecting watches.

CHARLES J. GARVEY, born on April 5, 1934 in New York City. Commissioned a 2nd lieutenant upon graduating from West Point in June 1956. Attended the Infantry Basic Officer Course and Airborne/Ranger schools. Assigned to the 25th Inf. Div. as an infantry platoon leader. Won the division Military Stakes competition and received the Gen. I.D. White Award. Also commanded the winning division drill team.

Assigned to The Old Guard in 1960. Worked in ops on the shows Torch Light Tattoo and Prelude to Taps. Promoted to captain and selected to command the Honor Guard Company. Attended the Armor Advanced Course in 1962.

Sent to Vietnam as an infantry battalion advisor, then as an advisor to the Commandant of Cadets at the Vietnam Military Academy.

Received a master's at Indiana University and taught at West Point for three years. Returned to Vietnam and served as a battalion executive officer in the 101st Airmobile Div. In 1970 he attended the Armed Forces Staff College, then commanded the 2/509 Airborne/Mechanized Battalion in Germany. In 1974 he graduated from the Army War College and received a master's from Shippensburg University. Assigned to the Office of the Deputy Chief of Staff for R D and A and then to the Office of the Under Secretary of Defense for R and E. During that time received a MBA from American University.

Retired in 1982. Awards include Bronze Stars, Air Medals, VN Gallantry Crosses w/Palm, Meritorious Service Medals, and the Legion of Merit.

Worked for two small firms as a vice president of software engineering. Retired in 1994. Married with four children and four grandchildren. Hobbies are investing, real estate and gardening. Lives in Vienna, VA.

RONALD D. GAZDA, Specialist 4th Class, Honor Guard Co., 1st Battle Group, 3rd Inf. He was born in Minneapolis, MN Aug. 29, 1940. Ronald enlisted in the US Army on Dec. 15, 1959 and completed basic training at Fort Leonard Wood, MO. He graduated first in his class in combat communications.

He was selected to Honor Guard Company in April 1960; was chosen as battle group "Soldier of the Month" in October 1961; and was promoted to Specialist 4th class.

His most memorable moments in Honor Guard Co. were participating in the inaugural ceremonies for President John F. Kennedy. While in Honor Guard Co. he met Anne Newman of South Portland, ME. They were married in April 1962. Ron and Anne have two daughters and four grandchildren. Ron was the CEO of a national trucking company and retired in 1990. He currently lives in Minnesota and Florida.

TIM GOLDEN, born Watertown, NY, 1960. Graduated St. Mary's of the Angels (grammar school), Ukiah, CA, 1975. Graduated Chico Senior High School in 1979, Chico, CA. Enlisted in the US Army Sept. 4, 1979. Basic/AIT training Fort Benning, GA, September to November 1979. Duty station Fort Myer, December 1979 to September 1982, Co. E - Honor Guard, 1st Marching Platoon, and 2nd Casket Platoon.

Highest rank was Spec 4. Discharged in September 1982. Attended Butte College January 1983

to December 1985, AA degree; Chico State University, Chico, January 1986 to 1989, MA degree, Recreation & Parks Management.

Employed as Park Ranger for the US Army Corps of Engineers, April 1991 to present, Englebright Lake, Smartville, CA.

WILLIAM R. GORE, LTC (Ret.), served in USNR from 1947-48 with duty onboard the USS *Small* (DD-838). Joined the Army and was assigned to 5th Inf. Div. Transferred to Co. E, 3rd Inf. Regt. at Fort McNair until 1950.

Released to 318th PIR, 80th Abn. Inf. Div. in 1950. Commissioned as 2nd lieutenant, Infantry, in 1952. Transferred to 2174th US Army Garrison in 1966 and retired as LTC AG in 1976.

Still serving in American Legion, Military Officers Assn., Employer Support of the Guard and Reserve and the 80th Div. Reserve Assn. He is also a proud life time member of The Old Guard Assn.

BENJAMIN F. GOTTFRIED, CW2 (Ret.) US Army, life member TOGA 0330. He was born in Philadelphia, PA and from there entered into US Army on March 26, 1959. His assignment to The Old Guard as the maintenance officer began in early 1976. Shortly after being assigned he was volunteered for the position of OIC of the motorcade for the inaugural parade of President Carter. The motor pool was also responsible for the jeep used in the annual Torch Light Tattoo Ceremony (Now called Spirit of America). The jeep was a replica of Gen. George Patton's and their motor sergeant dressed up as Patton and rode in the jeep during the ceremony.

They began having problems with the jeep that made it impossible to repair it to the standards they wanted. He made some phone calls and finally found a replacement jeep (M38A1) but it was located in a property disposal yard in Wisconsin. He made a phone call to the operations officer at Dover AFB and after giving him a story (Ben might have stretched the facts a little, which he won't go into here) he agreed to bring the jeep back on their next mission to Wisconsin. With some minor repairs and a paint job they were back in business with their new jeep!

His assignment with TOG was much too short. He was reassigned to Fort Belvoir in late 1977 to fill a critical vacant position at the Directorate of Maintenance until that position was re-designated as a civilian position. He was then transferred to the Brigade Maintenance, still at Fort Belvoir, where he remained until he retired after 20 years of service.

His 20 years included one tour of duty in France, two tours of duty in Germany and two tours in Vietnam, plus state side assignments to Texas, California, Virginia, New Jersey, South Carolina, Missouri and Oklahoma.

His personal medals are the Bronze Star, Purple Heart, Army Commendation, Good Conduct (3 awards) and Meritorious Service. Then there's the National Defense, Vietnam Service, Vietnam Campaign, Vietnam Cross of Gallantry w/ palm and the Unit Meritorious Service Medals. Through all of this his most memorable assignment was TOG.

DONALD "DON" J. GREEN, LTC USA (Ret.), born in Iowa, 1932; high school, Renwick, IA; BS degree and ROTC commission, University of Iowa, 1954; member varsity and ROTC rifle team, 1950-54, tied 23-year-old university indoor small bore record.

Military: 1955-58. BIOC w/PCS to the 33rd Inf. Regt., CZ, Panama; member, USARCARIB Rifle Team and invited to Olympic rifle marksmanship trials in 1957; 1958-60, Commo. Ofcr., The Old Guard; 1960-63, Inf. Ofcr. Career Crse. and instructor, USAIS; 1964, Bn. Adv. ARVN; 1965-67, Fort Leonard Wood, MO, Co. Cmdr., Bde. S3, Bn. Cmdr., and Leadership School Commandant; 1967, AFSC en route back to RVN. Arrived in RVN in the midst of 1968 Tet Counter Offensive; successive assignments as Bn. XO, Bn. Cmdr., 2nd Bn., 12th Inf., 25th Inf. Div. and Staff Officer, 1 Fld. Force; 1969-72, ODCSPER, DA, was primary DA action officer for training center Drill Sergeant Program; 1972-75, instructor, CGSC, Fort Leavenworth, KS. Retired as LTC, June 1975.

Awards and Decorations: CIB, EIB, Legion of Merit, BS w/OLC, ARCOM w/OLC, AM w/10 OLCs, DA GS ID Badge, MSM, VN Cross of Gallantry w/Palm and Gold Star, NDSM, VNCM and VSM.

Special Comments: "When I was first assigned to The Old Guard, I was completely unaware of its special nature of TOG and did not fully appreciate the privilege and honor until much later. I now regard that fortunate assignment as pivotal in my overall military career. Officers with whom I served during the 1958-60 period will for life rank at the very top of my list of dear friends.

Post-Military to Present: 1975-83, Owner-Broker, C-21 Real Estate franchise, Leavenworth, KS; 1984-2000: VP & Trust Ofcr., UMB, Bank, Kansas City, MO. Retired, March 2000. Volunteer at St. Joseph Health Care Center, Kansas City, MO; Volunteer at Kansas City Middle School a "Youth Friend" with focus on the English as Second Language program.

HARRY E. GRIESER, born Dec. 20, 1929 in Springfield, OH. Attended Springfield Schools and graduated high school in 1947.

Entered the Army in August 1948 at Fort Dix, NJ. Took basic training at South Post, Fort Meyer in Co. H. After basic he was assigned to C Co., 3rd Inf. Regt., 1st Bn. at Fort Myer, VA.

He met his wife Joan while serving there. She was assigned to the WAC Detachment at Fort Meyer, VA. They have been married 53 years and have two children and three grandchildren.

Harry was discharged in August 1951 as a SFC. He retired in 1991 as a machinist making helicopter transmissions.

WILLIAM C. GRIMES, born Jan. 20, 1945 in Ashland, KY. Enlisted in US Army in June 1964 at Fort Knox, KY. Served in Delta Co., 3rd Inf., 1965; Inf. OCS graduate, 1966; and completed six years active service to include Vietnam, 1968-69 and instructed at Special Warfare School, Fort Bragg, NC. Completed 30+ years, retiring May 24, 2002 at Col/06 with 100th DIV (IT) USAR. Awards include LOM, BS, AM, SFTab and Order of St. George. Earned EIB at Fort Myer as PFC/E3.

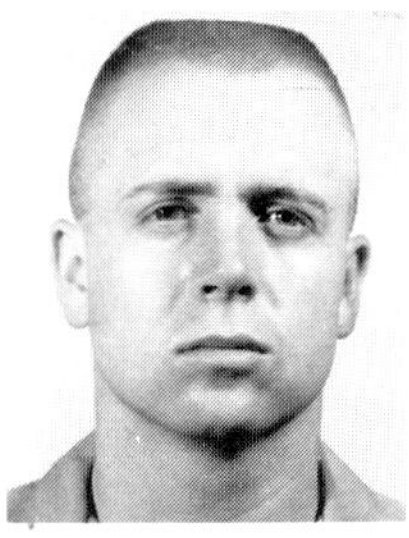

Married to Valorie Dobson Grimes, RN, Hospice Vistacare and have three sons (Bill, Matt and Jeffrey) and six grandchildren. Graduate of UL Law School (1975) and US Army War College (2000). Presently, Chief Deputy Prosecutor in Clark County, IN. Life member of Old Guard Association and VFW Post 1427. Gideon of 1st Baptist Church, Charlestown, IN.

JAMES D. GROOVER, joined the Army at the Knoxville, TN, AFEES Station on April 12, 1971 and sent to basic training at Fort Knox, KY, then on to AIT at Fort Polk, LA. From there he was assigned to The Old Guard at Arlington, VA with HHC Co.

He had several assignments at the White House and Capitol Bldg. and Andrew's AFB. He was in President Lyndon Johnson's Funeral, President Richard Nixon's return trip from China, and also President Nixon's 2nd inauguration riot control as a soldier with the 3rd Inf., The Old Guard.

After leaving the Army on April 11, 1973, he later became a police officer with the Knoxville, TN Police Dept. and eventually became a police motorcycle officer for Knoxville Police. He retired in June 1998 and now enjoys retirement, along with his twin sons, Nick and Matt. He still resides in Friendsville, TN.

ROBERT HANEY, born Nov. 12, 1945 in Cohoes, NY. He graduated from Schuylerville Central High School in June 1964 and enlisted in the US Army Sept. 9, 1964. Upon graduation from basic and advanced infantry training at Fort Dix, NJ, he was assigned to the 1st Bn. 3rd Inf. Regt. "The Old Guard," and stationed at Fort Myer, VA. He was in the Bravo Company's marching platoon and also the company guidon. He obtain the rank of sergeant on Sept. 16, 1966.

On Nov. 27, 1965 he married his high school sweetheart, Melodee A. Priest. Melodee and Bob are parents of a daughter, Tamee, and a son, Bob III.

Shortly after his separation from the Army, Bob and Melodee moved back to their home town of Schuylerville, NY where he has been involved in the local youth sports, the Cub Scouting program, and the Lions Club. On June 3, 1983 Bob had the honor of participating in a wreath laying ceremony at "The Tomb" through the Scouting program.

He received his bachelor of science degree from The College of St. Rose at Albany, NY and has worked for the General Electric Co., Fleet Bank and New York State's Office of Children and Family Services in the information technology field.

THOMAS HANSON, born Oct. 13, 1960 in Minneapolis, MN. He served as a fife instrumentalist in The Old Guard Fife and Drum Corps at Fort Myer, VA from 1979 to 1982. He was the volunteer Human Relations/Equal Opportunity representative for Co. H for two years. His rank upon separation was Specialist Sixth Class.

He is a 1987 graduate of the University of Minnesota where he earned a bachelor of arts degree in geography with supporting fields in planning and public policy from the Hubert H. Humphrey School of Public Affairs. In 1991 while in graduate school studying American cities, he was diagnosed with schizophrenia.

Today, he is the founder and editor of a monthly literary journal with a circulation of over 100 readers. In 2002, he was named Poet Laureate of The Old Guard Association by President Donald Cofsky. He lives in south Minneapolis.

CARL L. HARDY, born Jan. 4, 1947 in Concordia, KS. Enlisted September 1965 to August 1968, then October 1973 to December 1990. Enlisted in Santa Ana, CA and Fresno, CO; served with 1/61st Inf., 1/8th Inf., "TOG" 1975 to 1980; recruiting duty, 2/26th Inf., 3rd/8th and 5th/8th Inf.

Served in Germany twice and the family loved it. Carl received the ARCOM from "TOG" and Gold Recruiter Badge w/3 diamonds. Retired in Astoria, OR in 1990.

Met his wife Susan in college and have two children, Patrick and Julie, and five grandchildren. Pat served with 1st ID in 1991 during Desert Storm. Moving to Colorado Springs, CO to raise horses.

CHARLES W. HART, born Oct. 27, 1927. Entered service Sept. 28, 1950, went to Fort Dix, NJ, then to Fort Myer. He was assigned to A Co.

1st Bn. until his discharge Sept. 28, 1952.

He was married on Dec. 27, 1950 to Jacqueline who passed away in 1981. He married Gloria in 1984. Charles worked for Cons. Nat. Gas Co. from November 1952 until retirement in December 1987. He was a volunteer fireman for 45 years and belonged to the American Legion for 50 years.

He lives in Dillener, PA and has three children and four grandchildren.

JAMES T. HAWTHORNE, born Jan. 25, 1961 at Tampa General Hospital, Tampa, FL. Enlisted in US Army (delayed entry program) in 1978. After graduation from Brandon Senior High School in 1979, he attended basic Army Infantry School at Fort Benning, GA from July to October.

Reported to the 3rd US Inf. Div., The Old Guard, in October 1979. Assigned to Co. A, "Commander in Chief's Guard." Assigned to 3rd (casket) Platoon. Over the next three and a half years, he was promoted to SGT (E-5), team leader in 1st Squad, as well as First Casket Team leader.

Returned to hometown of Tampa in 1984 to pursue employment as a civilian. In March 2002, he married Robin, and in October 2002 they welcomed their daughter Emily into the world. She had a 10-year-old stepbrother, Matthew, from Robin's previous marriage. Currently, James is working for a trucking company in Tampa, FL.

RONALD L. HENRY, born Nov. 29, 1937 in Columbus, OH. Enlisted in the US Army in September 1967. Served in the US Air Force from April 1956 to March 1984. Other assignments include 2nd Inf. Div. in Korea, US Army Recruiting Command and two tours in Vietnam with the 1/5 Inf. (Mechanized) and the 11th ACR.

He retired from the US Army as a first sergeant in 1979. Ron was awarded the Soldiers Medal, Purple Heart, Meritorious Service Medal and the Army Commendation for Valor.

Ron has three children, Ronnica, Rod, John and six grandchildren: John A., Adam, Amy, Sarah, Mindi and Josh (all Henry's). He is currently the CEO of Lord Henry Enterprises, Inc., in Nashville, TN. Lord Henry Enterprises, Inc. is a large internet marketing company.

WILLIAM H. HENRY JR., joined The Old Guard in 1951 following service as a platoon leader and aide-de-camp in Korea. While Henry was "E" Co. commander from 1953-55, the company garnered every company level award in the regiment and set a record of more than 400 days without a reportable AWOL.

Designated as a Leavenworth honor graduate in 1963 and selected for the Army War College, Henry retired from the Army in 1967 as a lieutenant colonel. He entered civilian employment and eventually retired from Lockheed as corporate director of training and development. Qualified as an Army Aviator, Henry earned a BA from Ohio University, an MA from American University, and a doctorate from Rutgers University.

Married in 1952 in the Fort Myer Chapel, the Henrys are now retired in their ocean front home in Southern California where the enjoy entertaining their three children, nine grandchildren, relatives, and friends.

ROBERT L. HERSHEY, enlisted in the US Army on Sept. 30, 1948 and was sent to the 17th Abn. Div., Camp Picket, VA. After completing eight weeks of basic military training, he was sent to Fort Lesley J. McNair, Washington DC.

He was assigned to Co. E, 3rd Inf. and served until June 11, 1952. After serving a total of three years, eight months and 11 days of military service he was promoted to SFC on Feb. 1, 1952. He spent a lot of time training troops for the Korean War at A.P. Hill, VA. They all came back from Korea and said "You did a good job, sergeant; we all made it back home!"

JAMES R. HOPCUS, born July 17, 1944 in Oklahoma City, OK. He enlisted at, age 17 and reported to basic training at Fort Leonard Wood, MO on June 1, 1962. Several weeks into basic, he managed to injure a leg and was recycled. So he had the pleasure of spending the entire summer at Fort Leonard Wood.

He was sent to Fort Dix, NJ for AIT. Shortly before graduation several of them were called out of formation and sent to a meeting with an "Old Guard" representative, where they were offered a chance to defer whatever duty location they had chosen and become a member of The Old Guard. Only one guy out of 16 stood up and said he wanted to go to airborne training as planned and left.

It was a beautiful fall day when they arrived at Union Station in DC and what an impressive sight Washington is when you see it for the first time.

He eventually ended up in Charlie Company and became friends with Larry Free, Bob Weathers and Don Chapman, all cooks. At that time you could shift into another job and get on the job training. A job opened up in the mess hall and he took a cook's job, eventually having his MOS changed to 941.10 cook. He was a Spec. 4 when he was discharged on May 30, 1965.

His home is in Des Moines, IA. After 34 years, he retired as a field engineer from the local telephone company. Occasionally, he contracts back to the same company for a few months at a time.

Between them, James and his wife Jeanie have three children: Jason, Jordan and Tracey. No grandchildren yet.

JAMES A. HOVDA, 1SG (Ret.), born Jan. 23, 1939. He was assigned to Co. A, 3rd Bn., 3rd Inf., The Old Guard (TOG) in July 1983 as their first sergeant. During his 11 year plus tour of duty with TOG he was a 1SG in Alpha, Bravo, and HHC companies before retiring from TOG Aug. 15, 2004 and the US Army.

Enlisted in the US Army on Nov. 18, 1960 in Minneapolis, MN. Basic training was at Fort Leonard Wood, MO and communications training at Fort Knox, receiving MOS 051.10. Assigned to HHC, 42nd Atry, Chunchon, Korea. Left Korea after a 16 month tour and started Special Forces Training at Fort Benning, GA, completing Jump School in September 1962. Completed Special Forces training at Fort Bragg, NC, promoted to SGT E-5 and assigned to Co. D, team A/5-421.

James was sent to Vietnam in November 1962 and returned in May 1963. He was discharged from the Army in July 1964. Joined the USAR in March 1979 as a Senior Military Police Inst. (MOS 95B) with the 5042nd USAR School, Fort Snelling, MN. (He was employed as a police officer with the Minneapolis Police Dept.) In July 1983 he accepted assignment as the 1SG of Co. A, 3rd Bn., 3rd Inf. Regt., Fort Snelling, MN.

The 3rd Bn. of the 3rd US Inf. was the one reserve unit of the regiment. TOG's history in Minnesota and Fort Snelling, MN is long and prestigious. In 1994 the 205th IN Bde. Separate (L) was deactivated. The 3/3 IN now is part of our army history.

Awards/Medals: MSM (1 OLC), ARCOM (1 OLC), AAM (2 OLC), GCMDL, NDSM, AFEM, HSM, ASR, SS-M16A12, EX, 45, CIB, PRCHT BADGE, VNC of GWP (Unit) OS SVC BAR (1) and NPDR (4).

Jim Hovda is married to Jan and is honorably retired from the Minneapolis Police Department after 23 years of service. Jim lives in, Rice, MN and is active in the American Legion and other service organizations. Fishing and community service are his major projects. He has made over 600 sport parachute jumps.

JOHN HUME, born May 11, 1944 in Mexico City. Graduated University of Florida and commissioned as a 2LT Infantry USAR June 21, 1966. Completed ABN #20, IOBC #7 and JOC. Weapons Platoon Leader, "C" Co. (November 1967); Company Commander "E" Co. (February 1968) 2nd Bn., 3rd Inf.

Awarded Combat Infantryman Badge, Bronze Star Medal for meritorious service, Vietnam Campaign Medal and Valorous Unit Award. Promoted to Captain USAR in June 1972 and honorably discharged in November 1984.

Graduated University of Florida Law School, admitted to the Florida Bar in 1971, and currently practicing in Coral Springs, FL.

He served with many outstanding Old Guardsmen and wishes to especially remember LTC W.C. Carper, CPT Vesa Alakupi (KIA 1968), 1LT David K. Halen, SFC Charles H. Sandberg (KIA 1968) and SPC Alberto Caceres (KIA 1968).

ROBERT D. HURLEY, born Oct. 19, 1927, Jersey City, NJ and grew up in Teaneck, NJ. Served in the US Navy July 28, 1945 to Oct. 11, 1948 with duty aboard USS *Anikara* (ATF-98), Pacific Sea service, WWII.

Other Service: Gettysburg College, 1949-52, DMG; US Army ROTC, commissioned 2LT INF R Army; Infantry School, basic and advance; Armed Forces Staff College; Command and General Staff College; Korea, 1953, Co. A, 1st Bn., 19th Inf., 24th Div.; Berlin, Germany, 1961-64, Berlin Bde, Allied Staff Berlin DA, DSCOPS, Ops. Ctr.; Vietnam 1968-69, CO 2nd Bn., 505th Abn. Inf., 82nd Abn. Div., USA RVN; Fort Dix, NJ CO 3rd Bn., 1st Tng. Regt., USATC, deputy chief of staff USATC.

Awards include the CIB, LOM w/ OLC, BS w/V OLC, AM w/OLC, MSM, ARCOM w/4 OLCs, KSM, GSID w/BBS, VSM w/5 BBS.

Robert and Patricia have been married since April 14, 1956. They have a daughter Kiernan and a son Bobby.

ABDULLAH MUHAMMAD JOHNSON was assigned to the oldest active-duty infantry unit in the US Army. The 3rd US Inf. Regt. (The Old Guard) which has been serving our nation since 1784. The Old Guard is the Army's Official, Ceremonial Unit and Escort to the President. This unit is in the military district of Washington, DC, Fort Myer, VA.

Of the five specialty platoons in The Old Guard, Chief Johnson was assigned to the Caisson Platoon as platoon leader. This is the last active Equestrian unit in the US Army, and Chief Johnson has become the first African-American to hold this position in the history of The Old Guard and Caisson Platoon.

The US Army Caisson Platoon, currently housed in the barns that were built in 1908, came to Fort Myer, VA in 1948.

Chief Johnson is a native of the South Bronx, New York City and a graduate from the University of Maryland; he joined the Army in 1976 and was commissioned as a warrant officer in 1990. Chief Johnson has served in Desert Shield/

Storm and Operation Iraqi Freedom, completed three tours in Germany, two tours in South West Asia and numerous state side assignments. Chief Johnson is responsible for 48 infantryman (11B), one full-time farrier/blacksmith, one full-time leather man/saddle maker and 42 horses.

Chief Johnson arrived at The Old Guard June 1, 2004, and took over immediately with the coordination, selection of soldiers and horses and executing The Caisson for President Ronald Reagan's State Funeral.

A 28-year veteran, Chief Johnson still enjoys mentoring and training future leaders and being part of this important military tradition. *Submitted by a proud spouse, Jacque G. Johnson.*

NOTE: You can learn more about The Old Guard and the Caisson Platoon by visiting their website: www.fmmc.army.mil

HENRY W. JONES, born Feb. 5, 1917, Somerville, MA. Drafted in the Army in 1942, assigned to 3rd Inf. Regt. Antitank Co. Served in Europe from Lucky Strike to Darmstadt via Rennes, Wickrath, Babenhausen, Berlin.

Discharged in April 1946 with

rank SFC. Awards include the Good Conduct Medal, Victory Medal, EAME Theater Campaign and American Theater Campaign.

Married in September 1946 and went to California in 1947. He worked for Chevron Corp. until retirement in 1978. He has a son Steve, Retired Deputy Chief, Alameda Fire Dept., and two grandchildren. Henry belongs to VFW and the American Legion.

MICHAEL D. JUERS, born Feb. 23, 1955 in Wauwatosa, WI. After years of raising horses and working in the show horse industry as a manager/trainer, enlisted in the US Army in February 1976. Upon completion of basic training at Fort Dix, NJ and AIT at Fort Benning, GA, he requested and was assigned to The Old Guard Caisson Platoon 1976-80.

Married Yolonda Booth at Fort Myer Post Chapel 1977. They have one son, Nicholas.

During assignment to Caissons, he served as Senior Rider all positions, senior Caparisoned Horse Walker, Bravo - Fire TM. Ldr., Chief Saddler, Section Chief, White Horse Section, Duty NCO. Stables, Exercise and Barracks NCO.

Instrumental in design and development of Recruit and Remount Training Programs, facilities development and Saddle Shop reorganization.

Participated in numerous MDW Ceremonies, Battlefield Demonstrations, Pageants and more than 500 ANC Full Honors Funerals.

Recipient of Battalion's initial group awarding of The Old Guard Challenge Coin, Army Good Conduct Medal, and Army Commendation Medal.

Upon discharge, served with distinction: Falls Church, VA PD, Woodstock, IL PD and Illinois Army National Guard. Graduate Northern Virginia Criminal Justice Academy and Region 4 NCO Academy, Camp Ashland, NE.

Currently on staff at McHenry County College, Crystal Lake, IL. President/Owner, Lakeworth Landscape Environments, Inc. Life Member The Old Guard Association and North American Hunting Club. Now residing in Lake Geneva, WI.

BRAD A. KASPER, born in Wheeling, WV, Sept. 26, 1967. Basic training at Fort Benning, GA assigned to Harmony Church A-8-2. Selected for service in The Old Guard and assigned to Charlie Company, Casket (3rd) Platoon as a rifleman.

Served one year with Casket Platoon, then to Marching (1st) Platoon-Colors Squad for one year as RTO/ Team Leader for Charlie Company. Carried the Army's and was a guard on the color team.

During Charlie Company, selected battalion wide to serve as a Corporal Air Assault Instructor, Fort Belvoir, VA; selected and served two years as team leader and regular driller for the US Army Drill Team, Co. E (Honor Guard). Promoted to sergeant, became the senior catcher and head armour.

After leaving active duty in 1989 he served as a military policeman in the reserves for nine years in the 352nd MP Co., Oakdale, PA and 305th MP Co., Wheeling, WV. Promoted to staff sergeant and served as squad leader and platoon sergeant. Activated during Operation Desert Shield/Storm 1990/1991 with the 352nd MP Co. for 6-1/2 month assignment in Germany.

Left reserve duty in 1997 as a staff sergeant. Awards include; Expert Infantryman Badge, Air Assault Badge, Army Accommodation Medal (2nd OLC), Army Achievement Medal w/6 OLC, National Defense Service Medal, Overseas Ribbon with Numeral 1, NCO Professional Development Ribbon with Numeral 2, Physical Fitness Badge, Army Service Ribbon, Good Conduct Medal 1st Award, German Schutzen sneer (Silver).

During reserve service, he graduated from Ohio University (Cum Laude) with a BS in business, and Ohio Peace Officers Training Academy.

Worked as a policeman in Martins Ferry, OH for two years and served as vice president for local Fraternal Order of Police. United Parcel Service has employed him the past 14 years. Currently, as an operations supervisor prior as a pre-loader, air driver, package car driver and union steward for Teamsters.

Interests are military history, sports and physical fitness activities. Currently coaching scholastic wrestling at all levels and have for 15 years. Also involved as a coach in soccer and baseball.

Met and married Angie Carleton, his wife of 12 years. They have three boys named Gregory, Jacob and Drew. He hopes one of his sons will attend West Point and serve in The Old Guard. They all reside in Mansfield, OH.

RAYMOND E. KIRSCHENHEUTER, born Feb. 15, 1947 in Brooklyn, NY. Drafted in Army, July 1966 at Fort Jackson, SC. Served in Vietnam with 1st Cav. Div. Other assignments: Drill Sgt., Fort Dix, NJ; Honor Guard, Berlin Brigade, Germany; SFC (Tng. NCO) at Fort Leonard Wood, MO; 1SG of C Co., 3rd US Inf./(TOG), Fort Myer, VA.

While with The Old Guard, participated in many highly visible events such as both Inaugurations of Ronald Reagan, Gen. Omar Bradley's funeral, five Spirit of America shows, Torchlite Tatoo, Twilite Tatoo, and halftime show of Army-Navy Game, 1985 in Philadelphia, PA.

Ended his 21 years of service with 2nd Inf. Div., Korea. Awards received: MSM (2), ARCOM (5), AAM (4), Air Medal, PH, CIB, EIB, GCM (7). Retired in Willingboro, NJ.

Now working for US Postal Service. Member of TOGA, VFW, MOPH and volunteer for local Civil Emergency Response Team. Married to Phyllis since 1970.

HAROLD W. KRAMBECK, born Sept. 25, 1929, Moline, IL. He was a Civil Air Patrol Cadet, Moline, IL, 1946-47. In June 1948 graduated from United Township High School in East Moline, IL. Enlisted Aug. 4, 1948 in Army and discharged June 17, 1952 with the rank of corporal.

Basic training was at Fort Knox, KY, Aug. 4, 1948 to November 1948, assigned to 3rd Inf. Regt. Ceremonial A Co., Fort Myer, VA. North Post Services in Arlington National Cemetery and also family guide.

From May 14, 1949 to June 1952, Tomb of Unknown Soldier, walked the mat also had a relief changing the guard. Awarded Outstanding Performance of Duty while a member of the guard at the Tomb of the Unknown Soldier, Arlington National Cemetery. He has many cherished memories.

On June 10, 1950, he married Elizibeth Altomer at Fort Myer Chapel, have three sons. In June 1977 he married Viola Alger, has seven stepchildren, 26 grandchildren and two great-grandchildren. He worked 36 years at a match company as lead repairman and is now retired. He moved to Hudson, NY in 1952.

Harold is a member of American Legion, Hudson, NY.

ART KRISPIN, Major (Ret.), born in St Louis, MO on Oct. 13, 1948. He graduated with a BS in commerce from St. Louis University on June 12, 1971 and enlisted in the infantry on Aug. 2, 1971.

Active duty assignments were at Fort Leonard Wood, Fort Polk, Fort Myer, and Fort Benning. His platoon leader time was at Fort Lewis, WA. His AGR time was at Fort Meade, Newburgh, NY, and Fort Meade. He advanced in rank from private to major, retiring from the Army at Fort Myer on May 30, 1996. Currently he is employed in the air freight industry.

He was awarded the MSM (2nd award), ARCOM (2nd award), AAM, GCM, NDSM (2nd award), AFRM, EIB, Master Parachute Badge, Pathfinder Badge and Special Forces Tab.

He is a member of The Old Guard Association and is associated with the 11th SFG(A). He and his wife, Noreen, have two adult sons and live in Odenton, MD.

JONATHAN S. LACOTTI, Specialist 4, Co. H, First Presidential Salute Gun Battery. He was born Dec. 22, 1965, enlisted on Sept. 7, 1983 and attended OSUT for Indirect Fire Infantryman at Fort Benning, GA from September 1983 to December 1983. He was assigned to 3/66 Armor, 2nd Armd. Div., Fort Hood, TX from December 1983 to March 1986. He was assigned to Co. H, 3rd US Infantry (TOG), First Presidential Salute Gun Battery, Fort Myer, VA from April 1986 to July 1987.

He received one Army Commendation Medal, two Army Achievement Medals, and one Army Good Conduct Medal. He was discharged on July 6, 1987. He is a life member of The Old Guard Association.

He has been a police officer for over 15 years and currently resides in Perryville, MD with his wife and daughter.

ROGER A. LEAKE, SFC Ret., born in Harrisonburg, VA, Jan. 14, 1956, and currently resides in Radcliff, KY. Enlisted in the Virginia Army National Guard in June 1974 and graduated from infantry basic and AIT at Fort Polk, LA in November 1974.

Served as an infantryman with the National Guard until he enlisted for active duty in January 1979 and received AIT training as an administrative specialist at Fort Jackson, SC. Duty stations and assignments include Fort Knox, KY, administrative clerk; Fort Myers, VA (The Old Guard), Administrative NCO; Fort Shafter, HI, Administrative NCO; Fort Knox, KY, Personnel Tasking NCO; Louisville, KY, Army Recruiter; Riyadh, Saudi Arabia (Operation Desert Shield/Storm), Personnel Administrative NCOIC; Fort Knox, KY, NCOIC International Military Student Office and Instructor, US Armor School.

He met and married Rita K. Wathen, an employee with the Department of the Army. They are the parents of two sons and a daughter: Brian (SSG/US Army), Matthew

(SPC/KY Army National Guard) and Allison Marie. They also have two grandsons, Logan and Kyle.

He retired in July 1995 with the rank of sergeant first class, Fort Knox, KY. He held a dual status commission as Captain, USAR, Adjutant General Corps from September 1985 until his retirement. He received numerous awards and decorations that includes the Bronze Star Medal, Meritorious Service Medal w/2 OLCs, Army Commendation Medal w/6 OLCs, Army Achievement Medal w/3 OLCs, Army Good Conduct Medal (5th Award), National Defense Service Medal, Humanitarian Service Medal, Army Reserve Components Achievement Medal, Armed Forces Reserve Medal, Southwest Asia Service Medal w/3 Battle Stars, Kuwaiti Liberation Medals (Kuwait and Saudi Arabia), Korean Defense Service Medal, US Army Gold Recruiter Badge w/2 Sapphires, Expert Infantryman Badge and Meritorious Unit Citation.

He is currently working with the Department of Homeland Security, Immigration and Customs Enforcement in Louisville, KY.

DEAN R. LINDEMAN, born at the Patuxent River Naval Air Station, Lexington Park, MD on Jan. 24, 1956. Hometown: Rockford, IL. Enlisted in the US Army, Nov. 17, 1975. Completed basic training at Fort Dix, NJ and infantry AIT at Fort Polk, LA.

Assigned to the 3rd US Infantry in March 1976. Assigned to Co. C until November 1977, then transferred to the US Army Caisson Platoon. Honorably discharged in November 1980 as a sergeant. Member of Co. B, 2nd Bn., 129th Inf. IL ARNG, December 1980- December 1985. Honorable discharged as a staff sergeant.

Employed as a US Postal Service letter carrier from 1981 to present. Member of Northwest Fire Protection District, 1991 to present, currently serving in the rank of captain. Member of First Evangelical Free Church, Rockford, IL, currently serving as a member of the church council.

Married Nancy Lenney of Colton, NY in May 1980 and they have three children.

DONALD PAUL LIVINGSTON, born June 1, 1945, TOG July 1963 to September 1965. Enlisted March 29, 1963 Whithall ST, New York City. Bravo Company TOG casket team, firing party, funeral escorts, colonial musketeer in prelude to taps 1964/65, JFK funeral, Gen. MacArthur funeral, President Hoover funeral, LBJ Inauguration Parade.

Extended three year enlistment to volunteer to serve in South Vietnam with the First Infantry Division AOR III Corps, Alpha Co. 1st of the 18th Inf. tour September 1965-September 1966. Various action i.e. search and destroy, patrols, rung sat special zone operations. Assisting graves registration retrieve KIA special forces officers and men from the Michelin Rubber Plantation, WIA Nov. 14, 1965 while clearing DI AN for BRO HQ.

Discharged in September 1966. Awards include Purple Heart, CIB, VN Service Medal w/2 OLCs, National Defense Service Medal and Sharpshooter Badge.

His Army experience at the formative age of 17 through 21 set the foundation by direction and education for his successful career of 37 years in the aviation industry. Special thanks to Mrs. Smith Southpost Educational Center, who was instrumental in his future (wife of Col. Smith TOG HDQ).

Married March 7, 1970 to the former Mary M. Ferg of Johnstown, PA. Their son Capt. Michael Livingston is married to Janet and on active service in USMC. Donald is a life member MOPH, RNC, TOGA and member of FIRST ID SOCIETY, American Legion, VFW, 18th Infantry Association.

Retired from the position of Director of Ground Operations, Memphis, TN for Northwest Airlines after 37 years of service in various management positions in the carriers system. NWA management military official/liaison for Gulf War as well as Operation Iraqi Freedom accompanied troops to Saudi and Kuwait. Awarded Aerial Achievement Medal and Civilian Desert Shield and Desert Storm Medal.

Plans to travel and be active in local politics and enjoy the cabin in Virginia. Resides in Bridgetown, MS and Lexington, VA.

JOHN P. LOEHLEIN, born April 2, 1952 in DePere, WI. Enlisted in Army in March 1971 with basic training at Fort Campbell, KY. Advanced Infantry Training, Fort Polk, LA. Selected for The Old Guard in August 1971, a member of Co. E., "Honor Guard." Trained for, and was accepted as a Sentinel at The Tomb of the Unknown Soldier in November 1971. Remained at the Tomb until his discharge in March 1973.

Moved back to DePere. Married Janice VanDenElzen in May 1974. They have two children, Paul (born July 29, 1978) and Amy (born April 1, 1980). Divorced in 1992.

Employed at Paper Converting Machine Co. Green Bay, WI, starting in April 1973. Manufactured converting equipment for the paper industry.

Married Cynthia (Martin) Baeten on July 27, 2002, has two stepchildren, Melissa (born April 7, 1974) and Brent (born May 26, 1978).

He will always be grateful for the privilege of showing "Honor" to those who died in the service of our country.

EDWARD O. LOFLAND, born in Holyoke, MA, Nov. 10, 1925 and lived in Belchertown, MA. Enlisted in the Army Sept. 3, 1943. Called to active duty Nov. 26, 1943 at Fort Devens, MA. Transferred to Co. C,

3rd Inf. Regt. at Camp Butner, NC for basic training; then to Army Ground Forces Board #3, Fort Benning, GA.

Shipped overseas to northern France with The Old Guard and the 106th Inf. Div. Then to XXIII Corps. After re-enlisting with Co. C, lst Bn. 3rd Inf. Regt. at Limburg, Germany, he rotated to the States and the Infantry School Detachment at Fort Benning. He married his sweetheart, Martha E. Mabb in March 1946.

Served with the lst Inf. Div. in Europe, then to Camp Roberts, CA. Promoted from master sergeant to warrant officer. Then served at Camp Cooke and Fort MacArthur, CA. Transferred to 7th Inf. Div., Korea, then to Fort McClellan, AL and Fort Gordon, GA. Shipped overseas with assignment to Heidelberg Officers Club, Germany. Returned to Fort Devens, MA and the 29th Evac. Hosp. Retired in the grade of CWO-3 in November 1963.

With his wife, reared four boys: Ronald, Timothy, Michael and Matthew, and four girls: Patricia, Charlotte, Katherine and Martha. They have seven grandchildren. Retired, he continues to reside in Belchertown, MA with his wife of 58 years.

NICHOLAS D. MAJERUS, enlisted in the US Army from Minnesota and served one tour in Korea, assigned to HHC 1st Bn. (Reinforced) 3rd Inf., The Old Guard, 1963-64.

Memory: In the summer of 1963, they set up Red Carpet and Podium on the Whitehouse lawn. The Presidential Helicopter had never landed at the Whitehouse with a visiting dignitary, and they were doing a practice run. They had moved their trucks down in the trees on the road that circles the Whitehouse. As they were standing around their trucks waiting for the practice to be over, they saw the First Lady approaching them. Mrs. Kennedy was leading a pony with her daughter Caroline in the saddle. She stopped and talked to them for about 20 minutes until the helicopter had landed and then left. Nicholas will always remember how nice and pleasant she was to us common soldiers.

THEODORE PETER "TED" MANOLUKAS, born June 18, 1929 in Youngstown, OH. Drafted March 6, 1951 and sent to Fort Meade Separation Center. From there to Fort Myers along with 500 others to be sent to Korea as a regimental combat team after rigorous training.

Basic training was at Fort Myers, VA, Fort Mead, MD, Fort Belvoir and A.P. Hill. They were ready, but the Pentagon had sent orders – all enlisted men's rank was frozen and they were shipped to Korea according to how long they were on post. Hence, he was attached to A Co. of the 3rd Inf. until he was discharged March 9, 1953.

Ted was an Honor Guard at the Tomb of the Unknown Soldier from Oct. 17, 1951 to Oct. 27, 1952. He had the dubious distinction of being a passenger in a car driven by Big Joe Prieouxe when they hit the curb at the top of the hill in front of the COS (then Gen. Branderburg's house) and rolled over the cliff and into the ammo dump below. Fortunately they both walked away.

He met and married Peggy Speleotis of Boston, MA on Jan. 20, 1957. His wife is deceased. He has two sons, Nicholas; daughter-in-law Nancy; and Dr. John (PHd-DVM); and four grandchildren: Matt, TJ, Sara and Julia.

Ted retired in 1992 when he and his brother sold their business after 40 years. Today he lives alone in Boardman, OH; he visits his family and keeps in touch with two old friends from his old outfit: Dick Garrett at Barberton, OH and Dale Hoffman, Finlay, OH. They lost Henry Travelli of Ravenna, OH. last year.

TOM MARCOTTE, served with B Co. 2/3 from December 1968 to December 1969. He was a squad leader and later a SSG platoon leader. He was drafted in 1968, went to basic at Fort Leonard Wood, AIT at Fort McClellan, NCOES and Jump School at Fort Benning.

Leaving active duty in December 1969 he worked in the telephone industry for 35 years. In semi retirement he works in a high school library in Fishers, IN, where he runs the AV equipment, sponsors a Military History Club and is assistant coach of the Rugby Squad.

In 1970 he married and they have two children, who are both teachers, and four grandchildren.

In 1971 he joined the 12th Special Forces (reserve) and served for 14 years, later he was director of enlisted instruction in the 5030th USARF School. Promoted to CSM he was with the 300th CSB and Brigade CSM of the 8801st, retiring with 31 years service.

With B Co. 2/3 he was awarded a CIB, Silver Star, Bronze Star, and Purple Heart. At retirement he was awarded the Legion of Merit.

Currently he lives with his wife in Noblesville, IN. His only regret is not being young enough to do it all over again.

WILLIAM D. MASSIE, Supply Sergeant, Anti-Tank CO., 3rd Inf. Regt., born in Columbus, OH May 10, 1927, drafted and inducted into the Army July 23, 1945 at Ft Hays in Columbus, OH. Received basic training and also Intelligence and Recon training at Camp Fannin, TX.

Sent to Camp Pickett, VA and outfitted for overseas duty. Shipped out of New York on the USS *Lehigh Victory* and landed at LeHavre, France in January 1946. At a replacement center in Amberg, Germany he was assigned to the Anti-tank Co. 3rd Inf. Regt., then on duty in Ludwigsburg, Germany.

In Spring of 1946 the 3rd Regt. moved to Berlin to take over the American Sector occupation duties. Antitank Co. He was assigned the task of escorting prisoners and witnesses. The company also transported Displaced Persons to a point near their former homes. Those people were excited to be going home after years of forced labor in Germany. The one trip that he made like that, they delivered 25 boxcars of DP's and eight boxcars of PWs to Marburg, Germany.

In late November 1946, he left Germany from Bremmerhaven on the USS *General Stewart.* They docked in New York and were transported to Camp Dix where he was discharged Jan. 1, 1947.

He was not in the service very long, however, he is very happy and proud to have been a member of The Old Guard and is always eager to tell of its history.

On July 2, 1949 he married Jane Snider, his high school girlfriend and this year they celebrated their 55th year anniversary. They have two daughters: Jill, who has twin sons, Kyle and Steve, and a second daughter, Toni, who has one son, Nik. William and Jane are both retired and in 1990 they made the move from Ohio to Florida. They live in a golfing, retirement community and are able to play golf two or three time's a week.

FRANK L. MATHIEU, born Nov. 30, 1931, in Ellwood City, PA. In August 1949, he enlisted in the US Army, received basic training with 9th Inf. Div., Fort Dix, NJ. From November 1949 to August 1951, he was assigned to A Co., 3rd Inf. Regt., The Old Guard, Fort Myer, VA.

In August 1951 he served in Korea with H Co., 453rd Engr. Bn., truck driver through December 1951 and was with the 165th MPEG Co., 94th Military Police Battalion, MPEG through May 1952. Discharged in June 1952, Indiantown Gap Military Reservation, PA.

Married the former Nancy Reinheimer of Ellwood City, PA in 1957. They are the parents of two daughters and one son. They also have four grandchildren.

Frank graduated from Pennsylvania State University in 1961 with an associate degree in engineering.

He retired as the drafting/design department manager for Morrison Knudsen Corp. in 1995. His hobbies include hunting, fishing, golfing, and traveling. He resides in Dansville, NY with his wife, Nancy.

DONALD A. MAY, born in Buffalo, NY on Oct. 20, 1932. In March 1953 he was drafted and after basic training and Leadership School at Fort Dix, he was selected for The Old Guard at Fort Myer. Besides parades, cordons, guard duty, funerals and training, he became the armorer of Co. D.

They had many combat veterans from WWII and Korea. Their regimental commander was Col. Moore, his platoon leader had the Medal of Honor from WWII and his first sergeant was in the Bataan Death March.

Some of his memories include standing at parade rest in Arlington Cemetery at Gen. Vandenberg's funeral and 10 feet away was movie actor Jimmy Stewart, who was also a USAFR officer, talking to a group of senior officers. He also remembers playing catch with a sergeant just back from Korea. They had quit because the hot sun was drawing bits of shrapnel out of his back.

Donald was discharged in March 1955 with the rank of corporal. Since then he married Lucille and they have four children: Don, Donna, Dan and Dave, and nine grandchildren. He retired from New York Telephone Co. (Verizon) in April 1988. They still live in the Buffalo area and do a lot of traveling, and he spends a lot of time fishing.

They did not realize it then, but the time in The Old Guard was exciting and for some the best of times.

WILLIAM H. MCBROOM, born June 24, 1937, volunteered for the draft in 1955. Tagged with the nickname "Mac," after induction at Fort Campbell, KY, basic training at Fort Leonard Wood, MO, and advanced infantry training at Fort Hood, he was in the first cohort from there to join The Old Guard. Their predecessors had mostly been from Fort Dix.

The transition was accompanied by regional epithets between newcomers and old timers involving "Yankee" and "Rebel." McBroom was part of Fort McNair's 2nd Bn., George Co. from May 1956 until 1957.

Upon mustering out he attended The Ohio State University, earning a Ph.D. in 1968. He joined the faculty of The University of Montana in Missoula that year and retired there in 2004. He and his wife of 44 years, the former Gretchen Greiner, have two living sons (a third was killed in a head on crash in 1985). They have eight grandchildren.

JOHN A. MCCASLIN, born in Only, TN on Oct. 2, 1936. He was drafted into the Army in November 1959. During basic training he was selected to join the Third Infantry, First Battle Group, "The Old Guard." He was sent to Fort Myer, VA to train with the casket bearers at Arlington National Cemetery in Arlington, VA. His service with The Old Guard, from April 1960 until February 1962, consisted mainly of dignitary arrivals and departures, parades, funerals at Arlington National Cemetery, and other ceremonial duties.

During his time of service he had many memorable experiences such as participating in the inaugural parade for President John F. Kennedy, attending President Kennedy's Inaugural Ball, holding an umbrella over the head of Emperor Hirohito of Japan, opening the door for the Queen of Thailand, and modeling President Teddy Roosevelt's rough rider uniform on NBC's *Today* show.

In 1961, he married Cynthia Dailey. They are the proud parents of the following children: Barry McCaslin, Sherry McCaslin Peek, Myra McCaslin Gannon and Amy McCaslin Branch. They have two grandchildren, Ashton and Brittany Peek.

After his discharge, John began work with E.I. DuPont in New Johnsonville, TN. He was employed there for 37 years until his retirement in 1998. Since that time, he has enjoyed his retirement by spending time with his family, woodworking, and playing golf. He resides in McEwen, TN.

JOHN F. MCCULLOUGH, born Nov. 9, 1925. Enlisted Feb. 17, 1943, Philadelphia, PA and discharged June 30, 1954, Fort Devens, MA. His introduction to the 3rd Inf. Regt. occurred at Fort L.J. McNair in 1948 as a MP. They were called to formation in front of the building that would become the barracks for F Co., 2nd Bn., 3rd Inf. Regt.

They were informed they would be replaced by the 3rd Inf. Regt. They had the choice of transferring and remaining MPs or stay and become infantry if they met the criteria. To stay, they had to be between 5'10" to 6'2", not wear glasses, not be overweight or have facial scars. John was accepted and his career and life changed.

During the year of duty at Fort McNair, he had been court-martialed, acquired disabilities and met a wonderful young lady. He thought rejection was a real possibility. During the next four years, he went from private to master sergeant, got married, received numerous commendations and attended the Fort Benning Infantry school.

All of these memories can be attributed to great leadership from some great officers and sergeants, plus the many young soldiers that joined their ranks. Life would have been complete if his career had ended with the 3rd Inf. Regt. Korea called and he joined the second oldest active regiment, 9th Inf. Regt., 2nd Inf. Div. where he was awarded the Bronze Star Medal and many more fine officers and men to serve with. Upon return from Korea, he was sent to Fort Devens, MA. All his efforts to return to the 3rd Inf. Regt. were rejected and he resigned.

As a civilian, he served three terms on the Soldiers Relief Comm., two terms Michigan Veterans Trust Fund, member of the American Legion, Veterans of Foreign Wars, Old Guard Assn. and Disabled American Veterans where he has been elected to many local, state and national positions. Married 55 years, he has three fine children, six grandchildren and three great-grandchildren.

PETER J. MCDERMOTT, CW4 (Ret.), born in Astoria, Long Island, NY on Oct. 5, 1940. He enlisted in the Army on July 8, 1959 and after completing Initial Entry Training with the 2nd Inf. Div., Fort Benning, GA, he was assigned to Mortar Battery (Honor Guard) 1st BG, 3rd Inf. (The Old Guard) on Jan. 4, 1960.

During his tenure with The Old Guard, he was one of the initial members of The Old Guard Fife and Drum Corps and remained with them until his ETS on July 6, 1962.

Upon completion of active duty, he volunteered for Special Forces duty in the US Army Reserve and subsequently completed basic airborne training, the Special Forces Qualification Course, Jumpmaster School, the Intelligence Analyst Course, the Army Precommission Course, and the Psychological Operations Course before being appointed as a Warrant Officer in 1974. He continued to serve in the USAR with the 11th Special Forces Group (Abn.), the 352nd Civil Affairs Command, the 97th US Army Reserve Command, and the 311th Theater Signal Command. During Desert Shield/Storm, he served as the Chief, Reserve Personnel Division, Fort Meade Military Personnel Office for one year. He retired from the USAR on Oct. 5, 2002, as a CW4, after completing 43 years of military service. He had previously retired from civil service as a Department of the Army Civilian employee in October 1999 after 37 years of civil service duty.

CW4 (Ret.) McDermott has been awarded the Legion of Merit, Meritorious Service Medal w/2 OLCs, Army Commendation Medal w/3 OLCs, Master Parachutist Badge, Expert Infantryman's Badge, Special Forces Tab, British and Canadian Parachutist Wings, Good Conduct Medal, Armed Forces Expeditionary Medal, Army Reserve Components Achievement Medal (7th Award), Armed Forces Reserve Medal (4th Award), the NCO Professional Development Ribbon and the Army Service Ribbon. Additionally, he was awarded the Meritorious Unit Commendation as his unit was cited for Desert Shield/Storm duty and the Commander's Award for Civilian Service and the Superior Civilian Service Medal for his tenure as a Department of Army civilian employee.

He is married to Lorraine Weimorts McDermott and has one daughter, Kimberley Sean. He resides in Arnold, MD and is the 2004 President of The Old Guard Association.

JOHN J. MCGLUE, born and reared in Newark, NJ. Enlisted in the New Jersey National Guard in October 1947. Joined the Army in January 1949 and was sent to Fort Myer for basic training with the 3rd Inf. Regt. (The Old Guard). After 18

weeks of training he was sent to Co. F in Fort McNair. There the company performed MP duty in DC as well as normal 3rd Inf. duty.

When the Korean War started June 25, 1950, he joined 14 other volunteers from Co. F and 30 others from E and G Co., and they were sent to the 2nd Inf. Div. in Fort Lewis, WA. Three of them were sent to the 23rd Inf. Regt. and the other 42 men to the 38th Inf. Regt. John served with Co. D at the Pusan Perimeter, Kunu-Ri, Chipyong-Ni, and the spring campaigns.

He was awarded the CIB, two Purple Hearts, Korean Service Medal w/5 Battle Stars, UN Medal, American Defense Medal, two American and two Korean Presidential Unit Citations and others.

Returned to the States in July 1951 and was sent to Fort Benning, GA as an instructor at the Infantry School. After a short time he was sent back to Co. F at Fort McNair. He met his wife Jackie in DC and was discharged as a staff sergeant in August 1952.

They returned to the New Jersey shore area and reared two girls, Michelle and Lynne, and son, Tim. He worked as an electrician in construction and at the Pabst and Budweiser Brewery for 40 years. Retired in 1992 and now spends time walking on the Boardwalk in Belmar or playing golf. He also spends winters in Naples, FL, doing much of the same.

RICHARD "MAC" MCQUARRIE, born in Perth Amboy, NJ. He volunteered for the draft in 1961. His basic training started on December 4 at Fort Dix. He also attended the Trainee Leadership School and Advanced Infantry. After graduation ceremonies were over, his CO called Mac and five others to his office. They had all gone to the Trainee Leadership School. He told them that they were picked to go to Virginia and be in The Old Guard at Fort Meyer. Mac had previously been given orders for Frankfurt Germany and said he wanted to go to Germany. He said he did not want it on his record that he turned down this wonderful opportunity. He was right.

Off they went to Virginia. After Fort Meyer they were sent to A Co. at Fort McNair. At Fort McNair they performed many duties: flag folding, firing squad, cordon for the President and other dignitaries, and wreath laying ceremonies at Arlington Cemetery, as well as the Prelude to Tap Show. He was assigned to security for the White House during the march on Washington DC and was in the cordon for the JFK funeral. He is very proud to have served in The Old Guard.

Mac and his wife Bonnie have been married 38 years and live in Flemington, NJ. He has been an optician for 43 years and will soon retire and plans to move to Collegeville, PA. They have three children: Rick, Ryan and Shannon.

God Bless America!

FRANK T. MERRILL JR., entered the US Army in January 1970 with training at Fort Jackson in Columbia, SC. During training he took the tests for Officer's Training School and was accepted. However, President Nixon announced a cut-back in officers and therefore he was put on a delay. He was told to report to a meeting with an officer from Washington DC who offered him the opportunity to enter The Old Guard because of the cut-back at OCS and because he was "tall and slim." Frank gladly and proudly accepted.

Memories and high points in his service with "Charlie Guard" at Fort Myer include: marching in the funeral of Audie Leon Murphy in 1971; being honored as the "Soldier of the Month" for September 1971 (presented by Col. M.E. Lee); riot control during "May Day" demonstrations in DC in May 1971 (demonstrations were in protest of the bombing in Laos - four days in the Executive Office Building); half time at the Redskins games; half time at an Army game; Honor Cordons at the White House for visiting dignitaries-President Nixon always came out early and "chatted;" and "Torchlight Tattoo," wearing the Colonial wig and carrying the flag.

The discipline instilled in him as a member of The Old Guard has stood him well in his life and business career. He would take nothing for the experiences of his time in service. He was honorably discharged with the rank of E-5.

RICHARD G. MOELLER, born Jan. 12, 1933, Buffalo, NY. He was inducted Jan. 13, 1953 at Buffalo, NY, and took his basic training at Indiantown Gap Military Reservation, PA, January to May 1953.

Selected for duty at Fort Myer, VA as a member of Co. A, 3rd Inf. Regt. in May 1953. Duties included funerals and ceremonies in Arlington National Cemetery, parades and other military functions; some involving President Dwight D. Eisenhower and/or other foreign dignitaries. About November 1953, he became a member of the original Co. A, 3rd Inf., Old Guard Drill Team. The Drill Team performed intricate close-order drill exhibitions in and around the Washington, DC area as well as taking part in the regular Co. A everyday operations.

He was honorably discharged in January 1955.

HOMER R. MOSS, CSM (Ret.), born in Smithville, TN on Sept. 28, 1926. He enlisted in the Army on Jan. 12, 1945 and received his basic training with the Infantry Replacement Training Center at Camp Wolters, TX.

His other assignments include: 25th Inf. Div., Korea; 3rd Inf. Div., Germany; 3rd Inf. Regt. "The Old Guard," 1949, 1953 and 1958; 1st Armd. Div., Fort Hood, TX 25th Inf. Div., Hawaii; and Americal Division, Vietnam. His last assignment was command sergeant major of the 1st Cav. Div., Fort Hood, TX. He retired Jan. 31, 1975 at Fort Hood, TX.

He served a combat tour in Korea and Vietnam where he received his second award of the Combat Infantryman's Badge, the Silver Star, Bronze Star for Valor and the Legion of Merit. He married the former Addie Bell Donaghey of Pryor, OK. They are the parents of two sons and two grandsons. He makes his home in Killeen, TX. While serving in The Old Guard, he was chosen to represent the Army in a dedication ceremony celebration NATO's 3rd Anniversary. In attendance were President Harry Truman and Queen Juliana of Holland. He was also chosen to model the M-14 rifle for the Chief of Staff of the Army, Gen. Lyman Lemnitzer which was covered by *Parade Magazine*.

JOHN R. MURPHY, Sergeant, Co. G, 2nd Bn., 3rd Inf. Regt. "The Old Guard," 1952-54 at Fort McNair, Washington DC, after serving with Co. C, 14th Inf., 25th Inf. Div. in the Korean War, 1951-52 at the "Iron Triangle," Heartbreak Ridge" and the "Punch Bowl."

After discharge he graduated from Illinois State University with bachelor and master's degrees. He was a teacher and basketball coach for over 40 years. He had over 530 coaching victories and was inducted into Illinois Basketball coaches Hall of Fame.

John and his wife Jo Ellen have three children and four grandchildren. They live in Pontiac, IL. John has remained close to his friend, Bob Atchison, of Massapequa, NY.

CHARLES P. MURRAY, entered the Army from Wilmington, NC, in 1942, attended Infantry OCS and was commissioned 2nd lieutenant in 1943. He served during WWII in France, Germany and Austria with 3rd Inf. Div. His final combat assignment was as brigade commander in Vietnam, where he served in the 196th Light Inf. Bde. and 9th Inf. Div.

His awards include the Medal of Honor, Silver Star w/3 OLCs, Legion of Merit w/3 OCLs, Bronze Star w/ OLC, Air Medal w/6 OLCs, Purple Heart, French Legion of Honor and Croix de Guerre, and various RVN Commendation and Service medals.

He attended National War College and has degrees from University of North Carolina and George Washington University.

Col. Murray retired after 30 years active service at Fort Jackson in 1973 and lives with his wife in Columbia, SC. They have one daughter and two sons, both of whom are Vietnam veterans.

DALE A. NEWLIN JR., born in Philadelphia, PA, June 11, 1948 to US Army Retired Col. Dale A. Newlin Sr. and his wife. He enlisted in the Army in October 1979, finished infantry training at Fort Benning and was assigned to the 3rd US Infantry from 1980-83. A member of the Honor Guard, E. Co., for many historic ceremonies, he also served as driver and aide to two battalion commanders, Col. Don Phillips and Col. Robert Krause.

For many years he was Director of Admissions for Midstate Business College in Peoria, IL. For 14 years in San Jose, CA, he was an investment advisor for several banks. In 1995, he authored the book, *"Baseball's Greatest Careers."* Retired, he enjoys his hobby, roses, in Mount Dora, FL. He is a member of The Society for American Baseball Research, American Rose Society and The Old Guard Assn.

JOHN W. NICHOLSON, Brigadier General, born Feb. 22, 1934 in Struble, IA. He is a 1956 graduate of the US Military Academy, West Point, NY, where he received the Gen. MacArthur Leadership Award for his class. He is an airborne ranger combat infantryman. He served 2-1/2 years with infantry units in Vietnam. Other overseas assignments were in Germany, Korea, Lebanon and Switzerland. Served in The Old Guard 1960-62, platoon leader, Honor Guard, Company Commander.

Following 30 years active duty, BG Nicholson retired from the US Army. He has received "Several US and foreign awards and decorations for service and valor: Distinguished Service Medal, Defense Superior Service Medal, Legion of Merit w/OLC, Meritorious Service Medal, Bronze Star Medal for Valor w/4 OLCs, Air Medal w/V and 12 OLCs, Combat Infantry Badge, Ranger and Airborne.

BG Nicholson was nominated by President George W. Bush to serve as Under Secretary for Memorial Affairs in the Department of Veterans Affairs on Jan. 9, 2003 and confirmed by the Senate on April 11, 2003.

As Under Secretary, General Nicholson directs the National Cemetery Administration, a system of 120 national cemeteries providing dignified burial services for military veterans and eligible family members. He is responsible for land acquisition, cemetery construction, operation of the 120 national cemeteries, and for maintaining them as national shrines.

Gen. Nicholson also oversees other memorial related programs for veterans, including providing headstones and markers, administering federal grant programs to establish state veterans cemeteries, and issuing Presidential Memorial Certificates to survivors of honorably discharged veterans.

During his military career, Gen. Nicholson taught strategy and tactics at West Point, served on both the Army Staff and the Joint Chiefs of Staff at the Pentagon, and commanded the US Army Training Support Center. Gen. Nicholson served as a President Ronald Reagan appointee to Strategic Arms negotiations in Geneva, Switzerland, from 1982-84.

Civil Activity/Vets Organization: President Emeritus, International Society of the 173rd Abn. Bde.; Trustee, Virginia War Memorial Commission; Society of the Cincinnati; and Order of St. Hubertus. Gen. Nicholson holds a master's degree in public administration.

Married Sophie Sept. 1, 1974. They have seven children and 10 grandchildren.

THOMAS "GENE" NOOJIN, born Aug. 2, 1931 in Oneonta, AL, enlisted in the Alabama National Guard in September 1950. He served in the Korean War and was stationed at several locations within the United States and Germany.

In addition to the Korean War, he served two tours in Vietnam. During the Vietnam War, he served as a Vietnamese Ranger Advisor.

He received many honors and awards, including two Bronze Star Medals, Vietnam Service Medal,

Ranger Badge, National Defense Service Medal, 11th Air Assault Wings, Vietnamese Cross of Gallantry w/ Palm, ROK Presidential Unit Citation and Korean Service Medal.

Retired as master sergeant Oct. 1, 1970. Since 1965, he has owned and operated a residential rental business.

Married Josephine Zimmerman Jan. 15, 1954. They have two children, one grandchild and two great-grandchildren. They reside in Roanoke, VA.

Member of Colonial Baptist Church since 1953. Active as a Shriner and Mason. He is a charter member of 11th Air Assault Sky Soldiers.

EDWARD R. OKLESSON, born in Strabane, PA on Aug. 20, 1930. Currently retired from regular work and living in Inland Harbour near Reedville, VA. Enlisted on Aug. 1, 1948 and after induction at Fort Dix, NJ was assigned to South Post, Fort Myer, VA for basic training specifically for the 3rd Inf. Regt., "Old Guard" in Provisional Co. H.

After completion of 17 weeks basic training, he was assigned to "C" Co., 1st Bn. at North Post, Fort Myer, VA. Duties consisted of various ceremonial functions associated with burials of WWII Veterans in Arlington National Cemetery, Pentagon guard duty assignments, Precision Drill Team exercises, infantry training routines, and extra curricular activities on the Post Pistol Team.

After break-out of the Korean War, he was assigned as a drill sergeant to train recruits for the "new" Co. H. Discharged on Aug. 5, 1951 with the rank of sergeant.

Married Ima Jean Farrell "Gigi" on Oct. 27, 1951. They have two sons and four grandchildren. Civilian career followed as an industrial engineer after schooling at Naval Ordnance Laboratory and the University of Maryland.

Last assignment with the US Government was with the Naval Air Systems Command. Employed as an engineering consultant with LSA Inc. Arlington, VA until retirement in 1995. Currently involved in fishing, crabbing, oystering, enjoying Gigi's antique club activities, participating as a Docent at the Reedville Fisherman's Museum and American Legion Reedville Post 117 activities.

JOHN J. "JAKE" OKTAVEC, SGM (Ret.), born 1954 in Schonenberg, Germany. Entered military on July 17, 1972 from Connecticut. Completed BCT at Fort Dix. Assigned to The Old Guard Fife and Drum Corps as a Fife Instrumentalist Sept. 29, 1972. Positions included Assistant Fife Section Leader (1973-76), Fife Section Leader (1976-77), and Corps Sergeant (1977-82). Also served as the 1SG of H Co. from June to August 1981.

In November 1982 he assumed the responsibilities of the Corps SGM and was promoted to that grade in June 1983. Served as the regimental command sergeant major during the period April to June 1992. Awards include the LOM, MSM, ARCOM, AAM, GCMDL, NDSM, MUC, and the SUA.

In November 1973 he married Kathy Arrigoni, they have two daughters, Jessica and Erica. Jessica's husband Rob is currently serving in the Air Force. He retired from active military service Sept. 30, 2002 at Fort Myer, VA.

MICHAEL D. PERKINS, Captain, born in Walla Walla, WA on April 5, 1971. He enlisted in the Army in June 1989, attended basic training at Fort Benning, GA and reported to Fort Myer, VA on December 20. He served with "Battlin" Bravo Co. before trying out for the Continental Color Guard and moving to Echo "Honor Guard" Company.

His time in The Old Guard included field training exercises to Little Creek Amphibious Assault School, VA; Fort Stewart, GA; and a deployment to the National Training Center at Fort Irwin, CA. His ceremonies included missions carrying the National Colors at numerous State Dinners at the White House and carrying the Army Colors during the presentation of the battle streamers for The Defense of Saudi Arabia and The Liberation and Defense of Kuwait.

He graduated from the University of Idaho and was commissioned a second lieutenant in the Medical Service Corps in 1997. He served in various platoon leader positions and a company command in Honduras while at Fort Campbell, KY with the 101st Abn. Div.

He is currently serving as medical company commander for the 1st Bde. Cbt. Team, 1st Cav. Div. in support of Operation Iraqi Freedom II in Baghdad, Iraq.

He was married to the former Capt. Lillian Seini Close, an Army nurse, on Dec. 30, 2004 in Honolulu, Hawaii. They are planning on children.

His late grandfather, Vernon "Pete" Peterson, enlisted in the Army in Washington DC in 1943 and also served with The Old Guard before volunteering for the "Airborne" with the 82nd "All-American" Division.

NICHOLAS "NICK" J. PREVAS JR., born March 9, 1947, Baltimore, MD. Enlisted June 28, 1968, Fort Holabird, Baltimore, MD. Graduated from Infantry OCS, Fort Benning, GA, May 12, 1969. Served as platoon leader, 3rd Plt., Co. C, 4th Bn., 3rd Inf. Regt. (11th Lt. Inf. Bde., Americal Division), Central Highlands of Quang Ngai Provence, RVN, January through September 1970. Served as battalion assistant operations officer (S-3 Air), Fire Support Base San Juan Hill, October through December 1970.

Assigned to US Army Reserve April 21, 1971. Served as Intelligence and Operations Officer with 11th Special Forces Group (Airborne), Fort Meade, MD and as Force Development Training Officer with 310th Theater Army Area Command, Fort Belvoir, VA. Mobilized for Operations Desert Shield and Desert Storm: served as Strategic Intelligence Officer, US Army Intelligence and Threat Analysis Center, Washington, DC, November 1990 through July 1991.

Awards and Decorations: Bronze Star w/V Device and 2 OLCs, Purple Heart Medal, Meritorious Service Medal, Army Commendation Medal, Army Achievement Medal, National Defense Service Medal, Army Reserve Components Achievement Medal, Armed Forces Reserve Medal, Army Service Ribbon, RVN Service Medal, RVN Campaign Medal, Combat Infantry Badge and Parachute Badge.

Civilian Activity: Career Federal Civil Service, 1974-present. Currently serves as the Property Management Officer, Headquarters, US Marshals Service, Washington DC. Retired as Lieutenant Colonel, USAR in August 1993.

Nick and Carol Linda married July 23, 1966 and have two children, Virginia-Marie Prevas-Beckett and Nicholas J. Prevas III. Their first grandchild was born in March 2005. They live in Highland, MD.

LARRY J. REDMON, Major, is currently the Army Operations Officer, Joint Operations Division, Joint US Military Advisory Group-Thailand. He is responsible for planning and coordination of operations and exercises between the US Pacific Command and the Royal Thai Armed Forces, as well as interagency coordination with various US Government organizations represented in the US Country Team and with the Royal Thai Government. He also serves as the USCDRPACOM's primary counter-terrorism coordinator and liaison officer to the American Embassy in Bangkok.

Previous assignments: Commander, B 3/1st Special Forces Group (Airborne), Fort Lewis, WA. Exchange Officer to the Royal Thai Army while attending the Royal Thai Army Command and Staff College and later military advisor to the Thai Armed Forces Contingent in East Timor during INTERFET. Commander of Special Forces Operational Detachment Alpha 011 and 015, XO of A 1/10th Special Forces Group (Airborne), Germany. Platoon Leader and FDO, B 5/18th FA Regt., XO, D 3/30th FA Regt., Fort Sill, OK. Major Redmon also had 10 years enlisted service in various infantry and Special Forces assignments.

Major Redmon was born Sept. 1, 1961, in Logansport, IN, USA. He holds a BA degree in history from Cameron University, Lawton, OK. His military education includes: Royal Thai Army CGSC, US Army CGSC, Special Forces Officer Qualification Course, Infantry Officers Advanced Course, CAS3, FA Officers Basic Course, OCS, DLI, George C. Marshall Institute, Pathfinder, Jumpmaster, and Airborne schools.

Major Redmon is a Southeast Asian Foreign Area Officer (481). He speaks Thai and is rated in German and Russian. His awards include: the MSM w/OLC, JSCMDL, ARCOM w/ 2 OLCs, JSAM, AAM w/3 OLCs, AGCMDL (2 Awd.), NDSM (2 Awd.), AFEM, AFSM, HSM, NATO Medal, the Philippine Armed Forces Bronze Cross Medal, Australian INTERFET Medal, the EIB, SF Tab, Master Parachutist Badge, Pathfinder Badge, Tomb of the Unknown Soldier Identification Badge, Royal Thai Army Command and Staff Badge, Thai Master Parachutist Badge and numerous other foreign awards.

Major Redmon is married to Nuntalux Nakakositsakul, and they have son Khunsig, daughter Yiwhaa, son Christian and daughter Sara.

NORBERT F. REICH, born in Youngstown, OH, Dec. 12, 1928. Enlisted in the Army in August 1948, stationed at Fort Myer, VA and was discharged in August 1951.

Other assignments: lifeguard, served as Cadrea giving basic training, photographic work for the Army, unveiled statue of Field Marshall Sir John Dill, Color Guard for Premier of the movie *Battleground,* in charge of casket bearers of the first man buried in Arlington Cemetery, Lt. Howell Thomas, Color Guard for Secretary of Defense. It was a privilege to be in the Honor Guard and to serve his country.

He met and married Nancy Molinari and they have three daughters: Kathy, Sharon and Karen, and seven grandchildren: Jonathan, Alizee, Jackie, E.J., Rebecca, Kaitlyn and Mikayla. They live in Pittsburgh, (Baldwin) PA.

Employed in the steel industry for 41 years as draftsman, technical advisor, production manager, and purchasing department buyer. Retired in 1993.

In 1985, Nancy and he became Shaklee Distributors of natural food supplements, personal care, skin care and make-up, cleaning products and water and air purification systems.

BRETT P. REISTAD, born in Flint, MI in October 1956, and currently resides in Manassas, VA with his wife Jessica (Sabatino) and their four children.

Enlisted in the US Army through Greenwich, CT on June 28, 1974. Attended basic training at Fort Knox, KY (A-18-5), and Infantry AIT at Fort Polk, LA (C-3-1). Trained as an Infantry Light Indirect Fire Crewman (11C MOS - mortarman). Assigned to the Salute Gun Platoon (Presidential Salute Battery), Co. H, 1st Bn. (Reinf.), 3rd US Infantry (The Old Guard), at Fort Myer, VA in November 1974.

Ceremonial duties included loader, gunner, counter, assistant watchman, watchman, and NCOIC. Tactical duties included squad leader, training NCO, loader, gunner, and driver. Notable ceremonies include Spirit of America, 1975 & 1976; Bicentennial 50-gun Salute to the Nation, July 4, 1976; Scottish Military Tattoo at Wolftrap; July 1976; 1812 Overture, 1976 & 1977; Presidential Inauguration (Jimmy Carter), January 1977; and funeral of former Vice President Humphrey, January 1978.

Achievements include H Co. Soldier of the Month, April 1976; Graduate of Primary Non-Commissioned Officer's Course, Fort Knox, KY, March 1977; Promotion to Sergeant (E-5), November 1977; and, Graduate of Infantry Indirect Fire Crewman NCO Basic Correspondence Course, March 1978.

Decorations include National Defense Service Medal, Good Conduct Medal, Army Commendation Medal, Cold War Service Certificate, and Expert Marksmanship Badge (Rifle).

Completed enlistment on June 27, 1978 and was honorably discharged in January 1980. Retired from the Fairfax County (VA) Police Department in October 2003 as a lieutenant, after 25 years of service. Now employed by the Regional Organized Crime Information Center, a federally sponsored law enforcement program. Graduate of Bluefield College (VA) with BS in criminal justice.

Actively involved in The Old Guard Association since its infancy, serving as a member of its executive board, as president in 2001, and as vice president in 2002. Honored as a Distinguished Member of the Regiment in September 2003. Preparing to serve as Commander of The American Legion, Department (State) of Virginia, in 2005-2006.

SIDNEY L. RENFRO, born Sept. 11, 1938, Whittier, CA. Entered Army in June 1959 and sent to Korea in January 1960. Assigned to 7th ID, 3rd Inf. at Camp Kaiser. Returned to States in April 1961.

Married, he has two sons and three grandchildren. He graduated college in 1976, worked 20 years for Los Angeles Police Dept. and 13 years for Department of Energy. He was a member of Coast Guard, California National Guard, California Army Reserves, Tennessee Army National Guard and Texas Army National Guard where he retired in 1996.

He is presently retired from all his jobs and lives in Prescott, AZ.

VINCENT P. RIORDAN JR., born May 17, 1937, Worcester, MA. He enlisted Oct. 17, 1957, Springfield, MA. His assignments include 1st Regimental Combat Team, Fort Devens, MA and 1st Battle Group, 3rd Inf., D Co., June 1958.

He was company drummer for two years and drummed troops onto Andrews AFB for Nikita Kruschev's arrival in the Summer of 1959. D Company performed cordon duties.

Discharged Oct. 18, 1960 with the rank Specialist 4. Awards include the Sharpshooter and Rifleman's Badge.

Vincent was employed with CIA for 35 years. Retired May 17, 1987, returned on contract in November 1987 and retired as GS-14 in 1996. He married Margaret Hines Oct. 11, 1958 and they live in Chester, VA. They have three children: Joan Ellen, Margaret Ann and Vincent P. III. Grandchildren are Cameron, Chad, Joshua, Megan, Vincent P. IV and Coleen.

EDWARD J. RILEY, Acting Sergeant/SP4, Co. A, 1st Bn. (Reinf.), 3rd Inf. (The Old Guard), Fort McNair, Washington, DC, born in Birmingham, AL on Jan. 23, 1947. After graduating from the University of Cincinnati in architecture, he was drafted into the Army on Nov. 9, 1971 and completed basic training at Fort Polk, LA. He reported to Fort Myer, VA in January 1972 for advanced infantry training at AP Hill, Virginia. He was then assigned to Fort McNair, Washington, DC.

He participated in J. Edgar Hoover's funeral, President LBJ's funeral, President Harry Truman's memorial service, President Nixon's inauguration, and numerous full honor funerals in Arlington National Cemetery.

He was discharged from active duty on Nov. 8, 1973. He married his wife Arlene in Cincinnati, OH in 1977. He is a registered architect in seven states and has lived in Oklahoma, California, and currently resides in Arlington, VA.

TIMOTHY A. ROBINSON, born in Auburn, ME on April 9, 1966. He enlisted in the Army in January 1985 and went to basic at Fort Jackson, SC and AIT at Fort Gordon, GA. Trained as a 31 Kilo, Combat Signaler. He was discharged in December 1985 as an E-4.

Married Jodii Hayes in May 1995 and first child born in September 2004. He attended Rochester Institute of Technology after leaving the Army and majored in printing management and sciences.

Currently working at a large printing plant as pressroom supervisor, where he has been employed for 14 years. His wife is a supervisor at a nation-wide medical billing agency.

His best memories are doing the half-time show at the 1985 Army-Navy game, S.O.A., flying in and repelling from helicopters, and all the great times in Georgetown. Current hobbies include playing golf and softball and collecting sports memorabilia. He is an avid Red Sox and Patriots fan.

He is living in Maine where he was always meant to be.

VIRGIL L. ROECKER, born in Genoa, OH on Feb. 16, 1930. He was inducted into the US Army on March 14, 1951 and took his basic training at Fort Meyer in Arlington, VA. After basic training he was transferred to Co. E at Fort McNair in Washington,

DC until his discharge in March 1953 as a corporal.

He married Betty Schlievert on July 17, 1951 in Ohio. She moved to Washington with him and was employed as a secretary in the Pentagon in the Ordnance Department. They are the parents of two daughters and a son. They have three teenaged grandsons.

Virgil worked for Libbey-Owens Ford for 12 years and then for AP Parts, making mufflers and tail pipes for 30 years, retiring in 1992. He has been Commander of Genoa Legion Post 324 for several years and works part-time as a driver for Luther Home of Mercy.

GARY D. ROGERS, born Feb. 23, 1939, in Fort Worth, TX. Drafted in December 1961 in Dallas, TX, and after basic and advanced infantry training at Fort Chaffee, AR, he was assigned to 1st Battle Group, 3rd Inf., The Old Guard, remaining for the duration of service.

Discharged from regular service in November 1963, with the rank of Sergeant E5 and entered the 49th Armd. Div. of the Texas National Guard, and was discharged from the Texas National Guard in December 1965 with the rank of Staff Sergeant E6.

He married Fran Nimon on June 26, 1965; they have two children, a daughter, Allison, and a son, Glenn, and two grandchildren, Stephen and Emily. He was employed by Bell Helicopter Textron for 35 years and retired in March 2001.

Interests include military history, gardening and travel. He and his wife divide their time between Texas and a vacation home in Arkansas.

PATRICK J. RYNN, born on July 7, 1945 in Brooklyn, NY. He lived in Brooklyn until 1950 when his family moved to Port Chester, NY. He graduated from Port Chester High School in 1963, and attended his 40th high school reunion in 2003. The year 1963 was the year of the Cuban Missile Crisis. This is when he decided to join the Army and help out his country.

Enlisted at Whitehall Street New York, NY and was stationed at Fort Dix, NJ. It was also the day before President Kennedy was shot. He took his basic training at Fort Dix, signed up for Airborne Unassigned and got cook as his duty. He took an eight-week Cook's School course at Fort Dix immediately following basic training, then was assigned to Aurora, CO at the Fitzsimons General Hospital as a young, inexperienced cook. He arrived in Colorado in April 1964 and really enjoyed the Aurora, Denver, CO, area.

In January 1965 he received orders to be sent to Korea to an artillery 8-inch tower and then self-propelled howitzer battalion. This was in a small camp north of the 38th Parallel.

After 13 months in Korea, he received orders for Fort Meyer, VA. He was assigned to the 3rd Inf. He reported to Headquarters Co. but was assigned to Delta Co. where a first cook was needed.

He was PFC while in Denver, then made SPC4 and SPC5 while in Korea. His rank was SPC5 when he arrived in Virginia. After three years in the military he left while at Fort Meyer in November 1966.

Patrick went to work for United Parcel Post but was laid off on Christmas Eve 1966. On the first workday of 1967, Jan. 3, he was hired by Hartford Electric Light Co., which became Connecticut Light & Power, which is part of Northeast Utilities. After 38 years he is still working there, but hopes to retire soon.

Married May 25, 1980 and divorced five years later. He has no children. He lives for golf and tries to play whenever he can grab the chance. Hopefully, he will be able to play more when he retires. He also likes to walk, read and watch UCONN basketball – men and women. He lives in Stamford, CT and is planning on attending Old Guard reunion in the near future.

TOM SCHROEDER, SSG, USA (Ret.), born in Canton, OH Nov. 22, 1963. He entered active service at Fort Jackson, SC; assignments include Fort Riley, KS with 1st ID (Big Red One) serving as the Division Tactical Medic. He went overseas to VII US Corps, Special Troops Battalion, Stuttgart, Germany, Serving as the Corps Tactical Medic.

In 1987, he was assigned to 3rd US Infantry (TOG), Fort Myer, VA, with many assignments from line medic to evacuation sergeant. He returned overseas to 3-21st Inf. Bn., 25th Inf. Div. (Light), Schofield Barracks, Hawaii, serving as an assistant platoon sergeant. Walter Reed Army Medical Center, Washington DC serving as a platoon sergeant for the executive suites.

His most memorable tour was as an air assault instructor during the Persian Gulf War. His most memorable accomplishment in TOG was becoming one of the first medics to become a Regimental Superdoc in The Old Guard. He was assigned to TOG from November 1987 to May 1992.

WALTER I. SHAW, born in DuBois, PA, March 16, 1923. Inducted into service at Indiantown Gap on Nov. 13, 1943. Sent to Camp Butner, NC, then to Fort Benning, GA assigned to F Co. of the 3rd Inf.

Departed for overseas duty March 8, 1945. Arrived at LeHavre, France, then on to Germany. When the war ended he was transferred to Headquarters Company and taught automotive mechanics in Darmstadt, Germany.

Life member of Veterans of Foreign Wars. He was awarded the Good Conduct Medal, Bronze Star and medal for participation in the Battle of Northern France.

Discharged in May 1946 and returned to his hometown. He spent 45 years in sales and service of Mack Trucks.

Married 62 years to June; they have two sons, four grandchildren and four great-grandchildren.

Retired in 1986 and moved to Sun City, AZ where he still lives. He had the opportunity in 1976 to tour Germany and visited the building in Darmstadt where he had taught during the war.

MANUEL J. SILVA, born in Livermore, CA in 1920, passed away Sept. 2, 1987 and is buried in Arlington National Cemetery. Enlisted in the Army in 1940, he often stated that he never had a draft card.

Assigned to the 15th Inf. Regt. Fort Lewis, WA, the 3rd Inf. moved to North Africa in 1942. Seriously wounded in 1943, he returned to Camp Shelby, MS with the 65th Inf. Div. The 65th was sent to Europe and remained until the war was over, and he was discharged from the army.

During his years out of the military, he did various jobs, i.e. rodeoing and working for the American Can Co. In 1949 he returned to active duty and the 15th Regt. which was assigned to Korea in 1951. Wounded again, he was sent to Okinawa but applied to return to his old unit in Korea after recovery. From the 15th he was selected to carry their colors in President Eisenhower's Inaugural Parade. This resulted in being stationed as 1SG of a company, The Old Guard, at Fort Myer, VA, and thereafter as a personal assistant to Secretary of Defense, Charles Wilson.

While at Old Guard he accompanied six congressmen, including J.P.S. Deveraux of the Bataan Death March, on fact finding mission to 22 advisory groups around the world, meeting such leaders as Chiang Kai-shek, President Nagib of Egypt and a private audience with Pope Pius XII.

Married Evelyn Newby and they are parents of two daughters, Victoria and Cecelia, and have two grandchildren, John and Jessica Galpin.

Served in Vietnam from 1967-68 as SGM of the 1st Bde., 1st Inf. Div. and chief instructor of RVN School.

Other assignments included: Chief Instructor ROTC, Valley Forge Military Academy, Wayne, PA; 1SG (E-8), A Co. 30th Inf., Schweinfurt Germany; Chief Instructor University of Richmond, VA; in 1969 promoted to CSM(E-9) as Regimental CSM 6th Cav., Fort Meade, MD; CSM for Test and Evaluation Command at Aberdeen Proving Grounds, MD; and CSM of 4th Cav., Schofield Barracks, HI.

Retired Aug. 31, 1974. His awards include Silver Star w/OLC, Bronze Star for Valor w/3 clusters, Purple Heart w/2 clusters and Combat Infantry Badge (3rd Awd.).

MAURICE D. SMITH, MSgt., USA (Ret.), born in Norristown, PA, Nov. 12, 1924. He entered the US Army April 7, 1943. Assigned to the 1st Inf. Div., he saw European service in England, the Normandy landings and the Rhineland. After attending NCO School in Bamberg, Germany with the 1st Inf. Div.; he was assigned to Co. A of the 27th Inf. Regt. (The Wolfhounds), 25th Inf. Div., at the outbreak of the Korean War. Fighting along Korea's Iron Triangle, he distinguished himself winning the Bronze Star.

From 1953 to 1961 MSgt. Smith proudly served with the 3rd Inf. Regt., The Old Guard. MSgt. Smith served as platoon sergeant in Companies B, C and D. MSgt. Smith continued his career at Fort Wainwright as 1st sergeant in a light helicopter company. He served at Fort Lee as instructor in the QMC School. Fort Sam Houston was his next assignment as training NCO for the reserve units.

MSgt. Smith completed his 22 year career at the 6th US Army Escort Unit in Oakland, CA, where he received numerous commendations for the manner in which he escorted casualties from Vietnam. He received the Bronze Star, Army Commendation Medal, American Defense Medal, six Good Conduct Medals, EAME Campaign Medal w/Bronze Star, WWII Victory Medal, Army of Occupation Medal w/Germany Clasp, Korean Service Medal w/2 Bronze Stars, ROK Presidential Unit Citation, Combat Infantry Badge, United Nations Service Medal and the National Defense Service Medal.

MSgt. Smith is interred at Arlington National Cemetery and is survived by his wife, Dorothy M. Smith, four children and six grandchildren.

FRANKIE SOLACE, born in Prosperity, WV, July 10, 1930. Enlisted in Army March 1, 1949 and received basic training at Fort Myer, VA.

Assigned to Ceremonial/A Co. 3rd Inf. Regt. Performed military funerals in Arlington Cemetery and other honor guard duties in the Military District of Washington.

Most memorable details: carried British flag for Princess Elizabeth at her Wreath Ceremony at Tomb of Unknown Soldier; casket bearer for Leslie J. Coffelt's funeral; Honor Guard for Field Marshall Montgomery at Gen. Bradley's house and at Washington National Airport for Gen. MacArthur.

Last assigned to Fort Ord, CA as platoon sergeant. He was discharged Sept. 19, 1952 with the rank of sergeant.

Married Irene Techak, June 6, 1959. They have four children: Michael, Timothy, Steven and Kelle, and seven grandchildren.

He retired March 31, 1993, after 30 years with the Dept. of Air Force, testing and repairing guidance systems for missiles and aircraft in Heath, OH.

Now resides at Hebron, OH and has been a life member of The Old Guard Assn. since 1998.

RAYMOND A. STEWART, enlisted in the Army in the Fall of 1936, just a couple of weeks after spending a year in the CCC's. He was assigned to Co. B, 2nd Bn. of the 3rd Inf., Fort Snelling, MN. He liked the Army and knew somewhat to expect. They had Army officers in charge of the CCC's. When he was turned to duty in early 1937, he mounted guard for the first time. There were 50 of them. The Officer of the Day said, "those bucking for orderly, step forward," about 10 stepped forward; Raymond didn't know what he was talking about. The officer reached over and pulled him out of the ranks, and after the inspection, he was picked as the orderly.

He was sent up to headquarters and had a chair in the hallway by Gen. Stone's office; he was the post commander. When his office closed for the day, he was sent back to the Guard Headquarters. He did not have to walk a post or chase prisoners. He was made orderly the next three times they mounted guard.

In the spring, the whole of the Fort walked to Fort Ripley, way over a hundred miles. They went by all the back roads. He was to report to the Motor Pool. They gave him a brand new 1937 Plymouth 4-door sedan and assigned him to the damaged property control. He was to chauffeur a major around, so while his whole company and all of Fort Snelling trudged along, and fought their athletes foot, Raymond drove along with his full pack in the trunk of the car.

Each night they camped and he shared a pup tent with his No. 4 in the rear rank. After they got to Fort Ripley, he would drive into Little Falls, the home of Lindberg, and pick up this major and take him to his office at the airport. Some days they would be gone all day settling claims where two cows were accidentally shot, or damaged some fences, or buzzed a mink farm and scared the minks into killing their young. Some days they didn't do anything, and he would visit with the pilots. There was a squadron of WWI Jenny's that practiced strafing and dive-bombing. When just the pilot went up, they would strap a sack of grain in the empty seat.

Raymond befriended a pilot and asked him if he could ride with him. He got permission from his supervisor. The next afternoon they strapped a chute on Raymond and lifted him into the Jenny and took off to about 10,000 feet, then into a dive to drop one of his bombs. Raymond knows he blacked out a couple of times during the dive. The bomb was made out of cardboard and filled with flour. When it hit, it busted and it looked like smoke. He leveled off, then flew around shooting at targets, then climbed back up and did the same thing over again. While he was strafing and started to climb up to do his third bomb dive, his motor started to cough. Raymond started to pray; he knew it was the end. The pilot put it in a tight turn back to the airport. Raymond swore then, he would never get into another airplane, although he later got a private license and its still valid at his age of 87.

Raymond might add this was the biggest peace time Army exercise ever held in the USA. In peacetime, there were observers from every major company in the world. They were practicing the pincher maneuver, and the Germans were especially interested. They all had clipboards and cameras. They later used that maneuver in most all their battles successfully.

He might add here that at Fort Snelling, they had the 14th Field Artillery, the 7th Tank Corps. Once a week or so, the 14th would take over the whole parade ground to practice. That was something to see. He had seen a lot of rodeos and chuck wagon races, but they couldn't hold a candle to the 14th; 25 to 30 rigs all full gallop, pulling cannon supply wagons, Caisson's Hospital wagons, big S's figure eights just missing death by inches, all going at a full gallop. He will never forget that sight. At this maneuver, they even had the 13th Cav. from the Dakotas, beautiful horses and good riders. They took charge of the communications delivering messages.

When it was all over, Raymond drove his Plymouth back to Fort Snelling (while his buddies walked) with his full pack in the trunk and no athletes foot. He was made PFC in the Fall of 1937. That was unusual as no one hardly made PFC in their first enlistment. Usually it was given as an incentive for re-enlistment. He could go on and on, but have to stop someplace.

RICHARD J. TESTRAKE, born on a farm near Ripley, NY, Sept. 14, 1938 and moved to Erie, PA in 1948. He joined Pennsylvania National Guard in 1955. Entered active duty with US Army in June 1957. Basic training was at Fort Knox, KY, with advanced infantry training at Fort Ord, CA. He served in Fort Knox, KY before joining "A" Co, 1st BG, 3rd Inf., The Old Guard in 1958.

While serving with The Old Guard, he participated in the interment of the Unknowns of WWII and the Korean Conflict at Arlington Cemetery, as well as the funeral services for former Secretary of State John Foster Dulles at Washington Cathedral.

Left The Old Guard in April 1960 for training as an Army Air Traffic Controller at Keesler AFB, MS. After graduating in November 1960, he was transferred to the EUSA, Korea, and served with the 1st Cav. Div. until transfer in March 1962 for duty at Davison Army Air Field, VA. He was honorably discharged from the Army on May 3, 1963 in the grade of E-5.

Married to Peggy since June 6, 1959. They are parents of one daughter whom they adopted in Korea as an infant, and one son stationed with the Coast Guard in Juneau, AK. They are grandparents of three and great-grandparents of one. Richard retired from GE as a welder in 2001 after 37 years. They reside in Erie, PA.

ROBERT MITCHELL THOMPSON, born in Manchester, MD on Sept. 4, 1939. He was drafted into the US Army on March 4, 1963 with basic training at Fort Jackson, SC and advanced infantry training at Fort Gordon, GA.

Robert was selected to go to Fort Myer, Arlington, VA, 3rd US Inf. (The

Old Guard). The Old Guard is the US Army official ceremonial unit and serves as escort to the President. Assigned to B Co., 3rd Bn. with rank E-4.

He was at the funeral for John F. Kennedy on Nov. 26, 1963. He was an Honor Guard on the steps of the capitol. He also guarded Kennedy's grave for about six months. He was at Gen. MacArthur's funeral, plus a lot of other funerals in Arlington Cemetery.

It was the best experience of his life to be in the Army at Fort Myer, Arlington, VA. He was discharged March 5, 1965 plus two years in the Army Reserve.

Married to Marlene Thompson (retired) and after 30 years as tractor-trailer driver, he retired. They live in Hampstead, MD. E-mail marthom@carr.org

THOMAS TWOMEY JR., CSM (Ret.), born in Washington, DC, Aug. 12, 1930. Enlisted in the Army Aug. 28, 1947, received his basic training with the 9th ID, Fort Dix, NJ.

Other assignments: 6th ID, Korea; 24th ID, Japan; 3rd ID, Fort Benning, GA; The Old Guard; l0lst Abn. Div., Fort Campbell, KY; 8th ID (ABN) Germany; 6th Inf. Regt., Berlin, Germany; 46th Co., Thailand, Special Forces; and 82nd Abn. Div. Fort Bragg, NC. He retired Jan. 1, 1978, in a ceremony at Comney Hall, Fort Myer, VA.

He served in both the Korean and Vietnam War, and was recalled to active duty in 1990-91, during Desert Storm. He married the former Erika Schulze of Berlin, Germany. They are parents of three sons and two daughters.

While serving in The Old Guard he was selected as the official Presidential Color Bearer for President Harry S. Truman and Dwight D. Eisenhower. He carried the colors during the 1953 inauguration.

In 1998 he was instrumental in organizing The Old Guard Association and was the first president of the association. He has the title of "President Emeritus."

WILLIAM R. "BILL" UNDERWOOD, Sergeant, born April 11, 1972. He enlisted in September 1989, Richmond, VA with the following assignments: 1990-94, 3rd US Infantry The Old Guard; 1995-96, 2nd Armd. Cav. Regt., Fort Polk, LA; 1996-97, 2nd Inf. Div., Camp Castle, Korea; 1997-98, MDW Engr. Co., Special Rescue, Fort Belvoir, VA; 1998-2000, 116th Inf., 29th Inf. Div., VANG Charlottesville, VA; September-December 2003, 2nd Bn., 317th Inf., USAR Lynchburg, VA; December 2003-present, Operation Iraqi Freedom 1/2 1st Cav. Div. Baghdad, Iraq.

Still serving, his medals so far include Purple Heart, Army Commendation Medal w/4 OLCs, Joint Service Achievement Medal, Army Achievement Medal w/5 OLCs, Reserve Achievement Medal w/3 OLCs, National Defense Medal w/2 Stars, Overseas Service, Reserve Component Overseas, War on Terrorism, War on Terrorism Expeditionary and Expert Infantry Badge.

Graduated Albemarle County Police Academy in 1995 and is currently employed as a patrol officer and crime scene investigator. He is also a member of Police Dept. Honor Guard and Surveillance and Observation Group.

Married Jennifer Jenkins on Oct. 5, 2002 and when he completes his tour in Iraq, they will make their home in Kesmick, VA. He is a member of Fraternal Order of Police, The Old Guard Assn., American Legion, Elks Lodge B.Poe #389, VFW, Virginia Firefighters Assn. and National Tactical Officers Assn.

CORNELIUS "DUTCH" VANDER STELT, born Jan. 23, 1934, Artesia, CA. Drafted Sept. 4, 1956 for two years active duty. Trained at Fort Carson, CO, 9th Inf. Div.

Assigned 1957-58 to Co. F and Co. A, 3rd Inf. Regt., Old Guard, Fort McNair, Washington DC. His Old Guard service highlights: the Queen and Prince of England visit ceremonies in 1957; Double Interment of WWII and Korean War Unknowns in 1958; Co. A trophy 1958, best Old Guard Marching Company at Fort Myer parades.

Served two years active Army Reserve at Long Beach, CA with Co C, 63rd Inf. Div. and served two years Standby Reserve. Honorably discharged Aug. 30, 1962 with rank SGT E-5 Team Leader.

Married Nettie Niehof March 20, 1959. They have five sons, one daughter and five grandchildren. Retired in 1994 from dairy farm and lives in Albany, OR.

ANTHONY J. VERNALI, born in Norfolk, CT March 3, 1933. He received basic training at Fort Dix, NJ, beginning on April 6, 1953. Upon completion he received orders to join the 3rd Inf. Regt. at Fort Myers, VA. During his service there with A Co., he had the honor of participating in many services at Arlington Cemetery. At that time A Co. housed the Army drill team, the Caisson Platoon, Cemetery ceremonial squads, and Sentinels of the Tomb of the Unknown Soldier.

After participating on the drill team and at many cemetery ceremonial duties, he had the special honor of becoming a Sentinel at the Tomb of the Unknown Soldier and remained there for the last seven months of his military duty.

After leaving the service he returned to work as a mason on many construction projects in Connecticut as a member of Local One Connecticut Bricklayers until his retirement in 1994.

He married his childhood sweetheart, Lucille. They have a son Anthony and a daughter Rosanne; also, five grandchildren: Brendan, Briana, Gabrielle, Benjamin and Patrick. They currently reside in Manchester, CT.

MARK D. VOORHIES, born 1946 in San Francisco. Conscripted in January 1968, trained in infantry and artillery and served in Old Guard (A Co., 4/3, 11LIB, Duc Pho, Vietnam) August 1968-May 1969. He was known as "The Professor" amongst his fellow grunts. Mark was wounded twice and profiled to Americal HQ until his discharge in August 1969.

He reports: "While forced to serve in Vietnam, I learned many lessons including respect for others, distrust of government and trust in my own judgement." He advises today's soldiers, "Be informed and objective, love yourself and others; leave a positive impact, but a small footprint."

After a GI Bill education, Mark served in Department of Veterans Affairs, Blind Rehabilitation Services in the west and southwestern regions for over 25 years. His career included patient care, professional education and research. He retired from federal service in 1996. Mark now lives in Lake Tahoe, NV and provides counseling to individuals and consulting services to organizations.

ROBERT M. WALLACE, born in February 1941, York, PA and enlisted in the Army on June 10, 1959. He took basic at Fort Hood, TX with the 2nd AD and AIT in artillery at Fort Sill, OK. He was sent to Munich, Germany in December 1959. There he served in Headquarters Battery, 34th FA where he became a wireman in commo platoon.

With activation of the 11th FA, he was assigned there in HQ Battery as a switchboard operator. After the construction of the Berlin Wall, he was extended there for 16 months.

Discharged in 1962, he re-enlisted in July 1962 to The Old Guard. Assigned to A Co. at Fort McNair, in commo. Participated in the funerals of J.F. Kennedy and General MacArthur. In 1964 he was sent to 25th Inf. Div. in Hawaii where he served with the 8th Arty. until July 1965. After discharge he was employed by Met Ed Electric Co., until retirement in 2001. While in service he was given a Good Conduct and several other medals.

Robert has three daughters. He is presently on the board of York Health Corp., has 40 years service to the American Legion and 23 years in 40 and 8. Presently a life member of The Old Guard Association. He has one surviving sister and is living in York, PA.

GEORGE HENRY WAPLE III, born Feb. 8, 1921 and enlisted in 3rd Horse Cav., Fort Myer, VA, Aug. 20, 1938. Assigned to 83rd ID in June 1942. Promoted to master sergeant at age 21 in November 1942.

As first sergeant of 331st Inf., 83rd ID, he fought through the Hedgerows, Hurtgen Forest, Battle of the Bulge and across Europe to the Elbe River to meet the Russians in May 1945. He served in all five European campaigns.

Returned to Fort Myer, VA in November 1945, he was assigned as first sergeant of the Ceremonial Detachment that performed all ceremonies in Arlington Cemetery, parades and Honor Guards in the Washington DC area. This also included duties at the White House, 1946-47 as Head Usher for affairs of state under President Truman. He was also President Truman's wreath bearer at The Tomb of Unknown Soldier memorials on each Memorial Day. On April 5, 1948, Ceremonial Detachment became part of A Co., 3rd Inf. He is a charter member of The Old Guard and The Old Guard Assn.

Commissioned 2nd lieutenant Jan. 12, 1952 by Gen. Omar Bradley in his Pentagon Office. Assigned in 1952 to 3rd Bn., 31st Inf. Regt. of 7th ID, in Korea. Combat experiences included participation in the Korean War's last battle where the enemy attacked their positions across the Chowan Valley.

Other assignments include Aide de Camp to Maj. Gen. Lionel C. McGarr; Advanced Infantry School, Fort Benning; G-4 Section, Military District of Washington; Corrections Officer and Athletic Director of the US Army Disciplinary Barracks.

Capt. Waple retired in June 1962 after 24 years of service. His decorations include the Combat Infantry Badge w/star, Bronze Star w/2 OLCs, Commendation Medal w/4 clusters, Good Conduct Medal and others.

He now resides near Fort Monmouth, NJ where he enjoys golf and is involved in family and community activity. He is the author of *Country Boy Gone Soldiering.*

JOHN M. WATTS, born Aug. 23, 1953 in San Antonio, TX. Entered the US Army in September 1976 and served in the 3rd Inf., The Old Guard, at Fort Myer, VA. He is a member of H Co. Salute Gun Platoon from January 1977 until discharge in August 1980.

He attended NCO School at Fort Knox, KY and NBC Class at Fort Meade, MD. Achieved rank of E5 SGT and awarded the EIB. In last year of service he was NBC NCO for H Co.

Employed by the city of Kerrville, TX in September 1981 as a firefighter. Later, he served as fire inspector and finally as an arson investigator and deputy fire marshall. Retired in August 2001 and now manages family ranch.

Married in 1986 to Linda K. Summer, a 30-year-employee of the Kerrville VA Hospital. No children.

DAVID A. WILLIAMS, SFC-E7, USA (Ret.), served in The Old Guard from 1960-63, 1965-67, 1968-69, 1970-72.

CHARLES WILLIS, Major (Ret.), born in Kinston, AL. April 7, 1929. Enlisted in the Army in April 1947, received basic training at Fort Jackson, SC. Other assignments include 703 MP Bn. Fort Myer, VA, 1947; 3rd Inf. Regt., The Old Guard, 1948-51, 1st Cav. Div., Korea and Japan 1951-52; 3rd Inf. Regt., The Old Guard, 1953-54; 6th Inf. Regt. Berlin, Germany, 1955-56; OCS Fort Benning, GA, 1956-57; Fort Bliss, TX, 1957-58; 25th Inf. Div. Hawaii, 1958-61; The School Bde., Fort Benning, GA 1961-63; Senior Advisor ARVN Inf. Bn. Vietnam 1964; and his last assignment was with the Joint Support Command, Fort Ritchie, MD, 1965-68. Retired with the rank of major in April 1968.

Served two combat tours, Korea and Vietnam. Received two awards of the Combat Infantryman's Badge. He is an original member of The Old Guard reactivation in April 1948.

Married to the former Madge Stern of Arlington, VA. They are proud parents of five daughters and reside in Waynesboro, PA. They have eight grandchildren.

DELBERT J. WOLCOTT, born May 30, 1930 in Kenmore, NY. Enlisted in the Army in Buffalo, NY on March 11, 1951. The two years of his Army career were spent in The Old Guard at Fort Meyer, VA. He was discharged March 11, 1953 as PFC.

He was called Jersey Joe Wolcott so many times that most of the guys didn't know his real first name.

In October 1954 he married Adelene LaScala. They have three children: James, JoAnn and John, who have blessed them with seven grandchildren: Nicholas, Rene, Amber, Jeremy, Matthew, Leah and Nathan.

His civilian career was with New York Telephone Co. as a lineman and cable splicer. He retired in March 1990. All of his spare-time is spent watching his grandkids in school activities and in sports. In 1957 they built a new house in Chaffee, NY, a small town south of Buffalo.

Memories of The Old Guard

President Eisenhower about to place a wreath at the Lincoln Memorial, February 12, 1954. Richard G. Moeller is the 3rd Honor Guard member down the steps on the right.

Opposite Top: *Carter Inauguration, 1977. Courtesy of Brett P. Reistad.*

Opposite Bottom: *Pvts. Varrialle and Guranie standing guard of General John J. Pershing as he lay in state, July 1948. Courtesy of George Waple.*

Left: *Fort Meade.*

Below: *1st Platoon, C Company pass in review at the Retirement Ceremony of Platoon Leader 2nd Lt. Hennigan, 1968. Courtesy of J.P. Ganio.*

Courtesy of Timothy A. Robinson.

LEFT: *John Boyle & Loy Brown at A.P. Hill. Courtesy of Nickie Miller.*

BELOW: *Major General Robert Yerks, Lieutenant Colonel Henry Watson and Commanding Colonel Robert Clark at Change of Command Ceremony, January 1976. Courtesy of Dean R. Lindeman.*

OPPOSITE TOP: *Graveside service for General Walker at Arlington National Cemetery, January 2, 1951. Courtesy of G.W. Comer.*

OPPOSITE BOTTOM: *Casket Bearers from Co. A. at Arlington National Cemetery, 1973. Pictured are Chuck Leucht, Bob Oldham, Gary Epperly, Ed Riley. Courtesy of Ed Riley.*

BELOW: *January 20, 1953, just prior to falling out for the Eisenhower Inauguration Parade. From left: Richard C. Craig, Harold Treible, Phillip Dolan. Courtesy of Richard Craig.*

BELOW BOTTOM: *Funeral of General Jacob Devers, October 1979. Courtesy of Michael D. Juers.*

OPPOSITE: *Acting Sergeant Ed Riley, Co. A., Arlington National Cemetery, 1973. Courtesy of Ed Riley.*

NAVY

Member Roster of The Old Guard

The Old Guard Roster

Edward Allen
George Allen
David Almeter
Kenneth Ammerman
Robert Anderson
Lowell Andreesen
Harmon Andrews
Philip Andrews
Silvio Annunziata
Mark Armstrong
Raymond Arsenault
Victor Arsiniega
Greg Askew
Robert Atchison
Keith Atkins, Sr.
James Aux
Richard Azzaro
Robert Baer
Tyler Bakkensen
Bernard Balint
Paul Ballou
Richard Barrere
John Barrett, Jr.
Tris Barry, Sr.
Robert Bartruff, Sr.
Gardner Bates
Henry Baum
Richard Becker
Thomas Bedient
Anthony Bellina
Michael Bergman
William Bergman
Robert Bernard, Sr.
F. Clifton Berry, Jr
Albert Bertaccini
Darren Besse
Edward Bethune
Raymond Bigler
Henry Billings
John Binkley III
Richard Bird
Ronald Bisi
Robert Bjorklund
Daniel Blake
Howard Blehm
Thomas Block
Scott Boatright
John Bolen
Earnest Bomhoff
Thomas Bone
Charles Bowens
Norman Bowles
Darin Bowman
James Boyd
John Boyle
Richard Brace
Garr Bradley
James Bradley
Del Brandt
Sanford Brenna
Clarence Brewer, Jr.
Roger Brewster
Robert Briseno
Susan Brockman
Marvin Broudy
Connie Brown
Joseph Brown
Larry Brown
Ralph Brown
Willard Brummitt
Epifanio Bruschetta
Tracy Bryant
George Bryant III
Robert Buckingham
Michael Buras
John Butchko
Michael Butrum
Aubrey Butts
Charles Cadmus, Jr.
Bruce Calcari
Robert Callahan
Carroll Campbell
Darren Campbell
Michael Candler
Melton Cannon
Michael Cantrell
Freddie Capps, Jr.
James Cardamon
Don Carmichael
William Carr, Jr.
Ralph Casado, Jr.
Martin Cassidy
Anthony Cerniglia
John Chandler
Robert Chaney
Donald Chappell
Alfred Charest
Charles Charleston
Robert Chavanne
Raymond Chevaux
Leo Childs
Robert Chilton
Dennis Choplin
Donald Christie
Ronald Chura
Gordon Churchill
John Cilli
Steven Cirko
David Clark
Edward Clark
Robert Clark
John Clinkenbeard
James Coffey
Donald Cofsky
Sharon Cofsky
Thomas Cohoon
Steven Colbert
Roy Cole
Thomas Cole
Travis Coley
Paul Collier
Gerald Comer
James Connor
Arthur Cooke
Stephen Corbett
Billy Cordell
Lloyd Cosby
Stephen Coston, Sr
Paul Cote
John Couig III
Gerard Counts
Oliver Craft
Richard Craig
Richard Cross
William Crosson
Darrell Crotts
Arthur Crowley
Donna Crowley
David Crumbacker
Andres Cruz-Buitron
George Culik
Roy Cummings
Thomas Cupples
Edmund Cymanski
Dean Daghita
William Daknis
Joseph Daly
Francis Daniel
Robert Daugherty
Christopher Davis
Michael Davis
William Davis
William De Freitas
Anthony De Paola
George Deady
Nicolos Deardeuff
Joseph DeLalla
Gordon Delaney
Rodger Dell
Charles Demoss
Virgil Dennis
James Derrick
Richard Detjen
Garry Dew
Gary Dickens
William Dickey
Dennis Dimity
Timothy Doherty
Francis Dolan
George Doll, Jr.
Joseph Donato
Ronald Dooley
Gordon Dorn
Grant Douglas
Edison Dowdy, Sr.
Charles Drake III
John Dreste
Robert Dreste
Davis Drew
Richard Duda
Walter Dueno
Steven Dunfee
Matthew Durham
Daniel Dyer
Joseph Dyer
Toby Easter
Donald Eckstein
Craig Eckwood
Daniel Eddinger
Doug Edins
Lawrence Edlund
Glenn Eickelberg, Jr.
Ronald Elliott
Alvin Ellis
Mark Elrod
James Endicott, Jr.
Kyle Engel
Irene England
William England
Thor Erickson
Charles Eubanks
Thomas Evanko
Richard Evans
Edwin Eveland
David Farrell
Kenneth Fay
Robert Finton
William Fitzpatrick
Elmer Foreman
Nelson Forsburg
James Fosdyck
Dennis Fosket
Terry Foster
Donald Francisco
Michael Frazier
Eumid Freeman
James Friedman
James Fritz
Malcolm Frost
Ralph Frost
Richard Fuisz
Howard Furtado
Jana Galloway
John Ganio
Joe Garcia
Gregory Gardner
Lewis Garrett
Richard Garrett
John Garrison
Charles Garvey
Lewis Gayner
Ronald Gazda
Robert Geary
George Gennaoui
Joseph Gerard
Victor Giacometti
Joseph Gibbons
Donald Gibson
Michael Giglio
Vernon Gillespie
Talmadge Gilley
Richard Gillis
Edward Giomi
Edward Giomi
Kurt Glass
Joseph Gochenauer
Peter Gola
James Gomes
John Gonzalez
Donald Good
William Gore
Forrest Gossett
Benjamin Gottfried
Albert Graves
Larry Gray
Terry Gray
Donald Green
Martin Green
Bruce Greenhalgh
Jeffrey Greer
Michael Greer
Harry Grieser
Clemmie Griffin
Daniel Grigas
William Grimes
Jim Groover
Edmund Guditus
Joseph Gurganus
Norbert Habermehl
Craig Hackett
Stephen Halley
Charles Hambrick
David Handley
Robert Haney
James Hans
Thomas Hanson
Carl Hardy
Lowell Harlan
Franklin Harris
Robert Harris
Michael Harrison, Sr.
Charles Hart
Joseph Hartmann
Robert Hartmann
John Haskell
G. Ronald Hastings
Randy Hauck
George Haupt
Lionel Hawkins
James Hawthorne
Margaret Hawthorne
Robert Hawthorne
Rudolph Haynie
William Hedges
Robert Heintzelman
Martin Heli
Robert Helms
James Hennessee
Edward Henry
Ronald Henry
William Henry, Jr.
Robert Hershey
Danzel Hickle, Jr.
George Hicks III
Donald Hilbert
Mickey Hinkley
William Hisey
Charles Hodges
Gerd Hoffmann
Michael Hojnaski
J.R. Pete Holder
John Holmes
Michael Holub
John Holwege
James Hopcus
Timothy Horgan
James Hovda
George Howard
John Howard
Edward Howard, Jr.
Peter Hubert
Albert Huddy
Victor Hugo, Jr.
John Hume
Harold Hunt
Todd Hunter
David Huntoon, Jr.
Robert Hurley
William Hydukovich
Michael Ivankovits
Galen Jackman
Larry Jackson
Mark Jagger
Gregory James, Jr.
Jerry Janicki
Franklin Jarvis
Michael Jesse
Ole Jessen-Klixbuli
Douglas Jett
Jeffery Johann
Charles Johnson
George Johnson
James Johnson
Lynn Johnson
Robert Johnson
Thomas Johnson
James Joiner
Henry Jones
Thomas Jordan
Richard Joslyn, Jr.

Michael Juers
John Juneiman
Brad Kasper
Clarence Kelley, Jr.
Edward Kennedy
John Kievit
Stan Kingrey
Daniel Kirby
Ray Kirschenheuter
Scott Kittle
Lawrence Knighton
Harold Knoche
George Koch
Walter Konicki
Paul Konyha, Jr.
Charles Kottich
Jason Kowalski
Peter Kozak
Frederick Kozlowski
Harold Krambeck
Robert Krause
Kenneth Krauss
Arthur Krispin, Jr.
Donald Kronberg
Ronald Krul
Michael Kruse
Shawn Kruse
Wayne Kuschel
Anthony LaCaprucia
Jonathan LaCotti
William Lancaster
David Larose
Gary Larson
James Laufenburg
John Lauterbach
Peter Lawson
Roger Leake
Richard Leaman
John Lecrone
Michael Leczo
Richard Lee
Lonny LeGrand
John Lemen
Arthur Lessard
Jan Lilly
Dean Lindeman
Norman Lipinczyk
Joseph Lis
Douglas Litavec
Jimmy Little
Gregory Littleton
Candice Livermore
Donald Livingston
Craig Lloyd
John Loehlein
Edward Lofland
John Lohmeyer
Clark Long
Thomas LoPresti
Harold Lorenz, Jr.
Charles Love
Joseph Love
William Love
Marshall Lumley
Rupert Lyles, Jr.
William MacCumbee
John Madigan III
Nicholas Majerus
Ivory Mallory
Michael Mann
Theodore Manolukas
George March
Thomas Marcotte
Gene Marsh
William Marshall
Ralph Martenson
Nicholas Martinez
John Maryman
William Massie
Frank Mathieu
Miguel Matos
Michael Mattoch
Donald May
Roy May
William McBroom
John McCaslin
Gina McCloskey
John McCullough
Peter McDermott
Freddie McDonald
Nathaniel McElroy
Roy McElwee
John McGlue
Ronald McGuff
John Mcintire
Eric McKee
Raymond McLane
James McLaughlin
Don McNaughten
Richard McQuarrie, Sr.
Rolf Meisels
Robert Menist
Arthur Merat
Frank Merrill, Jr.
Alfred Meyer
Frank Meyer
Randal Meyer
Robert Meyle
David Milder
Donald Miller
Jan Milles
Christopher Mitchell
Medwin Mitchell
Kenneth Moats
Richard Moeller
Howel Mole
John Monahan
Ray Monjeau
Putnam Monroe
Henry Moore
Larry Moore
Junius Moore, Jr.
Terrance Moran
Mary Morelli
Brad Morrill
Christopher Morris
William Morris
Frank Morrison
William Morrison
David Moshier
Homer Moss
Irving Munroe
Patrick Munroe
John Murphy
John Murphy
Charles Murray
Laurence Nagle, Jr.
Stephen Nash
David Nelson
Joseph Nelson
Stanley Nersten
Dale Newlin, Jr.
Christopher Newman
John Nicholson
Dan Nolan
Thomas Noojin
Larry Norick
Todd Nosal
Stephen O'Connor
James O'Donnell
Daniel O'Keefe
Paul O'Mary
Charles O'Toole
Glenn Oden
Eddie Odom
Louie Odom
Edward Oklesson
John Oktavec
Ronald Olsen
Bradley Olson
Keith Olson
Ronald Orem
Jed Osborn
George Osborne
George Otis
Robert Overpeck
George Owens
Knox Ownby
Jeanne Pace
Francis Pachler, Jr
Thomas Page
Larry Parrish
Melvin Pax
John Peapos
John Pearce
Carl Peed
Elmer Pendleton
Granville Pennington
Michael Perkins
Robbi Perna
Harvey Perrit, Jr
Arthur Perritt
Donald Perry
Stephen Perry
Robert Phifer
Richard Pierce
Todd Pinney
Alfred Plew
Thomas Plunkett
Billy Porter
Scott Porter
Merland Poston
Martin Potter
Walter Powell
Kenneth Preston
Nicholas Prevas, Jr.
Thomas Price
Fortunato Priore
Anthony Pusatere
Sandra Quaschnick
Harold Quigley
John Quirk
Charles Rammel
Donald Rathman
Clifford Redden
Walter Reed
Toney Reese
Norbert Reich
Jerome Reichlmayr
Philip Reid
James Reilly
Brett Reistad
Sidney Renfro
Alan Resh
James Reynolds
Richard Rice
Don Richburg
Lenn Richter
Edward Riley
Vincent Riordan, Jr.
Arthur Rispoli
Richard Rivier, Jr.
Randall Roberts
Robert Roberts
Timothy Robinson
Berry Robison
Robert Rodriguez
Thomas Roe
Virgil Roecker
Gary Rogers
William Rommal
Michael Runnals
Michael Russell
David Rutt
Patrick Rynn
John Sassaman
Robert Saville
Paul Sayers
David Sayles
Nancy Schado
Joseph Schipf
Tommy Schroeder
Wayne Schurman
Timothy Scott
Blake Shaffer
Christopher Sharpe
Walter Shaw
Robert Sherwood
Catherine Shirley-King
Loring Shultz, Jr.
Evelyn Silva
Mark Simon
Harold Sims
Maurice Sims
Tommy Skiens
Steven Skorich, Jr.
George Sluzenski
Danny Smith
Donald Smith
Dorothy Smith
Herman Smith
Mark Smith
William Smith
Samuel Smithers
Robert Smudrick
Charles Sniffin
Errol Snyder, Jr.
Frankie Solace
Dennis Sommer
Walter Soukup
Anthony Sowder
William Speanburg
James Spires
Barry Spring
Stephen St.Clair
Gregory Stadler
Howard Steele, Jr.
Neil Stewart
Raymond Stewart
Wendall Stewart
Paul Stier
Robert Stoddard
Howard Stone
Wilbur Stovall III
Michael Sullivan
Travis Sullivan
Erik Syvertsen
Cliff Tailague
William Tallman
Edward Tatum
Charles Taylor
Denzel Teel, Sr.
James Teller
Joseph Temerario
Robert Tennant
Richard Testrake
James Thaman
Michael Theus
Robert Thompson
Scott Thompson
Charles Thoren
Brett Thornberry
George Thurston
Shawn Tilley
John Torstveit
James Towns
Bobbi Townsend
James Trepoy
Henry Trivelli
Michael Trotter
James Turner
Robert Turpin, Jr.
Thomas Twomey
Wilfred Ufkes
William Underwood
John Urevich
Curtis Utpadel
John Van Loan
Cornelius VanderStelt
Michael Vandre
Terry Vangen
Chester VanHorn, Jr.
George VanLandingham, Sr.
Charles Vanskiver
Pasquale Varallo
Dominick Varrialle
Anthony Vernali
Thomas Vick
John Vogelpohl
Alex Von Plinsky III
Mark Voorhies
David Wagner
Montie Wagner
Robert Walker
Albert Wall
Robert Wallace
Norman Wampler
George Waple
Glenn Watson
Paul Watson
John Watts
Jerome Waystack
Michael Welsh
Vernon Whitmore
Robert Whyte
William Willette
David Williams
Edsal Williams
Scott Williams
Spencer Williams
Charles Willis
Harry Wimbrough
Francis Winterburg
Daryl Winters
Delbert Wolcott
Bart Womack
Jerold Wood
Morton Wood, Jr.
Robert Woodruff
Robert Yancy
Larry Yommer
Robert Yoos
Jerry Young
Kristopher Young
Barrie Zais
Ronald Zak
Theodore Zdeb
Thomas Zurenda

Index

Index does not include The Old Guard Member Roster, listed alphabetically within its respected section.

Half-time show at the Army/Navy game, 1985. Courtesy of Timothy A. Robinson.

Regimental Sgt. Major Harry Wimbrough observing his Regiment as they prepare for an AFHR.

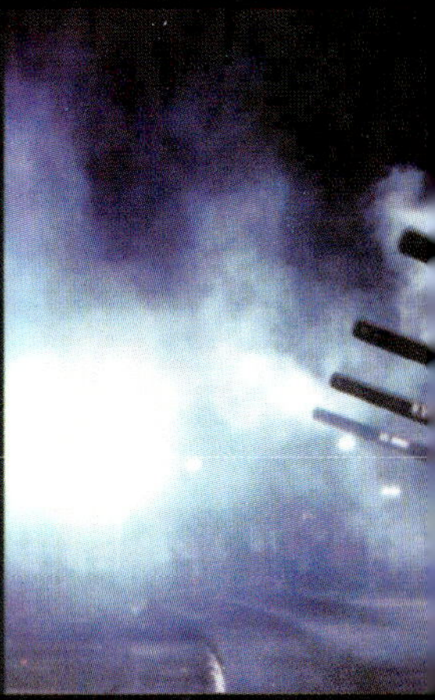